Advanced Mathematics

THEORY OF FUNCTIONS

OF

COMPLEX VARIABLES

By

Mohamed F. El-Hewie

TABLE OF CONTENTS

PREFACE

This is a revised edition of the chapter on Complex Variables, which was published few years ago in Part II of My Personal Study Notes in Advanced Mathematics. In this edition, I reproduced, refined, and enhanced all the **calculations and graphics** in a modern style of representation. In addition, I re-typed the cursive scripts of the personal notes and edited the typographic errors.

In the editing process, I added plenty of comments on the underlying meanings of the **arcane equations**, such that the reader could discern the practical weight of each mathematical formula. In a way, I attempted to convey a personal **sense and feeling** on the significance and philosophy of devising a mathematical equation that transcends into real-life emulation.

When equations deviate clearly from traditional algebraic patterns, I attempted to help the reader to understand how non-algebraic problems could be solved graphically and through programmed numerical iteration.

The reader will find this edition dense with graphic illustrations, made even for the simplest configurations that should spare the reader the trouble of searching other references in order to infer any missing steps. In my view, detailed graphic illustrations could sooth the harshness of arcane mathematical jargon, as well as expose the merits of the assumptions contemplated in the formulation.

In lieu of offering a dense textbook on Complex Variables, I opted to stick to my personal notes that give the memorable zest of a subject that could easily fade away in the mind when not frequently used

Mohamed F. El-Hewie

February 2, 2013

CHAPTER 1

INTRODUCTION TO COMPLEX NUMBERS

Complex numbers owe their genesis to the nature of the simple quadratic operation:

$$x^2 = c$$

where, **c** is a real number. Or, in more general form, the quadratic operation might contain a linear element, **bx** in addition to the quadratic element, ax^2, such that

$$ax^2 + bx + c = 0,$$

The roots of the above quadratic equation can be easily proven to take the quadratic formula

$$x = \frac{-b \pm \sqrt{b^2 - 4ac}}{2a},$$

In practice, x^2 could represent existing **area**, <u>added</u> area, or <u>removed</u> area on a surface. In the latter case, we get:

$$x^2 = -c$$

where, the negative sign signifies the **removal** of the area x^2 from the surface. If we assume that **x** is the side of a removed square, therefore, we get $x = \sqrt{-c}$. This is conventionally represented as $x = i\sqrt{c}$, where $i = \sqrt{-1}$. Similarly, the quadratic formula of the roots yields complex numbers if $4\,ac > b^2$.

Another way of looking at the genesis of the imaginary number, $i = \sqrt{-1}$, is **the squaring operation** of real numbers. Knowing that real numbers extend from $-\infty$ to $+\infty$, along the line of real numbers. Therefore, squaring implies the existence of an orthogonal axis to the line of the real numbers, such that the two axes, the line of real numbers and the orthogonal axis, form a plane of real and imaginary numbers.

Hence, a new field of mathematics of complex numbers is born with great utility in the solution of many algebraic equations, which cannot be solved in terms of real numbers.

Therefore, complex numbers extend the domain of real numbers so that basic algebraic operations can always be employed. Put differently, complex numbers comprise an **extension of the domain of real numbers**. Of course, we could do without complex numbers in solving algebraic equations, but not without great difficulties and cumbersome techniques.

The introduction of complex numbers and functions of a complex variable is commonly used in the following fields:

1. Integrating elementary functions.
2. Solving differential equations.
3. Electrical engineering.
4. Radio engineering.
5. Electrodynamics, hydrodynamics and aerodynamics.
6. Theory of elasticity.
7. Other natural sciences.

1.1. Algebraic properties of the complex number, $i = \sqrt{-1}$

The repetitive self- multiplication of $i = \sqrt{-1}$ results in the following interesting pattern.

$$i \ \ = i^5 = i^9 = i^{13} = \ldots\ldots = i^{4n+1} \qquad\ldots\ldots\ldots\ldots\ldots\ldots(1.1)$$
$$i^2 = -1 \ = i^6 = i^{10} = \ldots\ldots = i^{4n+2} \qquad\ldots\ldots\ldots\ldots\ldots\ldots(1.2)$$
$$i^3 = -i \ = i^7 = i^{11} = \ldots\ldots = i^{4n+3} \qquad\ldots\ldots\ldots\ldots\ldots\ldots(1.3)$$
$$i^4 = 1 \ = i^8 = i^{12} = \ldots\ldots = i^{4n+4} \qquad\ldots\ldots\ldots\ldots\ldots\ldots(1.4)$$

where, n = 0, 1, 2, ...

Example 1

Separate the real and imaginary parts of the complex function

$$z = (2x - i\,y)(x + i\,y)\,i$$

Solution

Upon distributing the three terms of z, we get

$$z = 2x^2 i - i^3 y^2 + i^2 x\,y$$

Using the multiplicative properties of $i = \sqrt{-1}$, equations (1.2) and (1.3), we get

$$z = 2x^2 i - i y^2 - x\,y$$

Hence, **z** could be written as

$$z = -x\,y + i\,(2x^2 - y^2)$$

In this final form, the complex variable z is clearly sorted in terms of its real and imaginary parts. Despite the fact that we made our example simple and direct, this type of algebraic processes arise in various ways during solving lengthy problems entailing multiple algebraic operations on complex variables.

7

Example 2

$$z = \frac{2}{2x - iy} - \frac{3}{x + iy} \quad ....(1.5)$$

Solution

In order to simplify the display of real and imaginary parts of z, we introduce the property of **conjugates of complex variables** as follows.

The conjugate complex variable of $(2x - i\,y)$ is: $2x + i\,y$ ....(1.6)

The **product of a complex variable by its conjugate** gives real quantity as follows.

$(2x - i\,y)(2x + i\,y) = 4x^2 + y^2$

$(x - i\,y)$ is the conjugate complex variable of $(x + i\,y)$

Similarly,

$(x + i\,y)(x - i\,y) = x^2 + y^2$ ....(1.7)

Therefore, z of equation (1.5) could be arranged by multiplying numerator and denominator of each of its terms by the conjugate of the denominator given in equation (1.6) to get:

$$z = \frac{2}{2x - iy} \cdot \frac{2x + iy}{2x + iy} - \frac{3}{x + iy} \cdot \frac{x - iy}{x - iy}$$

Using the properties of the products of conjugated complex variables we get

$$z = \frac{2(x + iy)}{4x^2 + y^2} - \frac{3(x - iy)}{x^2 + y^2}$$

We now could sort the real and imaginary terms as follows:

$$z = \left(\frac{2x}{4x^2 + y^2} - \frac{3x}{x^2 + y^2} \right) + i\left(\frac{2y}{4x^2 + y^2} + \frac{3y}{x^2 + y^2} \right)$$

CHAPTER 2

THEORY OF FUNCTIONS OF COMPLEX VARIABLES

2.1. The real and imaginary parts of a complex number

A complex number z is characterized by a pair of real numbers (a, b) having an established sequential order of the numbers a and b.

This is stated in the notation

$z (a, b)$.

The first number *a* of the pair (*a, b*) is called the **real part** of the complex number z and is denoted by the symbol

$a = Re\ z$.

The second number *b* of the pair (*a, b*) is called the **imaginary part** of the complex number z and is symbolized by

$b = Im\ z$.

1. Complex numbers (revisions)	$z (a, b)$
Any number in the form $x + i\,y$, where x and y are both real and $i = \sqrt{-1}$ is called a complex number. e.g., $6 - 4\,i$	z : complex number (a, b): pair of real numbers $a = Re\ z$: real part of the complex number z $b = Im\ z$: imaginary part of the complex number z

2.2. Complex number as vectors

Any complex number, $x + i\,y$, can be expressed by the alternative form,

$x + i\,y = r (\cos\theta + i\sin\theta)$.

$x + i\,y = r (\cos\theta + i\sin\theta)$ Therefore, $x = r\cos\theta$	$r = \sqrt{x^2 + y^2}$ $\cos\theta = \dfrac{x}{\sqrt{x^2 + y^2}}$

9

$y = r \sin \theta$ $x^2 + y^2 = r^2$	$\sin \theta = \dfrac{y}{\sqrt{x^2 + y^2}}$

2.3. Modulus and Argument of a complex number

$r = +\sqrt{x^2 + y^2}$ is called the "**modulus**" of the complex number $x + i\,y$. If we put

$z = x + i\,y$,

The **modulus z** is denoted by $|z|$,

The **angle** θ which satisfies the above two equations is called the "*argument*" or "*amplitude*" of the complex number.

If we keep the **angle** θ to satisfy in addition the inequality the $-\pi \le \theta \le \pi$, then θ is called the "***principle value***" of the argument or amplitude.

2.4. Cartesian and Polar representation of complex numbers

The complex number z,

$z = x + i\,y = r\,(\cos \theta + i \sin \theta)$

is represented on the **argand plane** by the Cartesian point $(x,\ y)$ or Polar point $(r,\ \theta)$.

2.5. Representation of Real and Imaginary numbers on the **argand plane**

The **argand plane** is an algebraic representation, not geometrical one, it serves the purpose of representing the **line of real numbers** across the **line of imaginary numbers**.

The **argand plane** if formed by the horizontal x-axis, where the **real numbers** are placed and a vertical y-axis, where the **imaginary numbers** are placed.

Thus, the complex number,

$z = x + i\,y = r\,(\cos \theta + i \sin \theta)$

10

is doubly represented, a polar vector of magnitude *r* and argument *θ,* and as a Cartesian paired components (*x, y*), as follows:

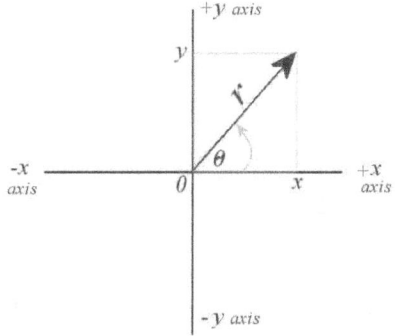

2.6. Representation of complex numbers on a **Riemann Sphere**

Let *L* be the complex argand pane of *z.* *A* is any point in the *L*-plane representing a complex number *z.*

Now, draw a sphere touching the plane *L* at point *0* of the intersection of *xy* axes. Let *SN* be the diameter of this sphere through *0*.

Join point *A* to point *N* with a straight line that intersects the sphere at *A'*.

Hence, for every complex number *A* in the z-plane, there is a corresponding point *A'* on the sphere.

If *A* is at infinity in the z-plane, then *NA* becomes parallel to the argand plane and *N* corresponds to all points at infinity. In the *z*-plane.

All the points in the plane *L* together with the points at infinity are said to inform the entire complex z-plane.

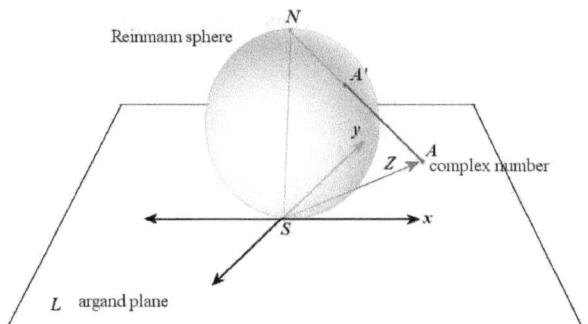

Two complex numbers z_1 (a_1 , b_1) and z_2 (a_2 , b_2) are equal only when both the real and imaginary parts are equal.

Thus, $z_1 = z_2$, only if, $a_1 = a_2$ and $b_1 = b_2$. It implies that, if $z_1 = z_2$, they both must share the same surface point on the Riemann Sphere.

Application of Riemann Transformation

Stringent conditions for utilizing Riemann surfaces are that the function in the z-plane must be:

1. Single-valued
2. Differentiable
3. Continuous

Each of the three conditions ensures that the 1-to-1 mapping from the $z(x,y)$-plane to the Riemann surface yield meaningful transformation.

We could map a line from the z-plane into a circle on the Riemann sphere, or a circle on the z-plane into a circle on the Riemann sphere. In each case, mapping is **one-to-one**, meaning each single point on the z-plane maps to a single point on the Riemann sphere.

The oldest application of such complex variable mapping dates to the first half of the nineteenth century. That is the Joukowski's mapping of a line and circle into the Riemann sphere that yielded the **airfoil theory**.

It is possible to relay on the series expansion of natural **logarithms** and **trigonometric functions** to reciprocate the conformal mapping between the z-plane and the Riemann sphere. Because the mapping functions are **transcendental, not algebraic**, we must truncate the expansions of the series to get first and second approximations of the **reciprocated transformation**.

Today, we do the reciprocation, or reverse mapping, between Riemann and z-plane with numerical iteration on fast computers. There are **no accurate solutions** for the airfoil theory through algebraic analysis alone.

Most universities and research labs develop their own computer algorithms to solve mapping transformation. We cannot reproduce the airfoil contours posted on all web sites such as the wikipedia, below, without possessing computational algorithm capable of solving transcendental equations. In MatLab, we must use **conditional loops** to find the roots of those notorious equations and our results depend on our selected accuracy window.

http://en.wikipedia.org/wiki/File:Wing_profile_nomenclature.svg
https://courses.cit.cornell.edu/mae3050/mae3050ThinAirfoils.pdf

Simply, Riemann transformation alters the [shape] of the curve due to the **inversion process**. It does not alter its [ordered] numerical contents. (Sort of folding a dollar bill onto different shapes without changing its monetary value. We could unfold the dollar bill and it still holds its initial value.)

Example 3

The best example of conformal mapping from Euclidean surface to Riemann Sphere is the Earth Globe reconstructed from the flat Earth map.

The flat Earth map cannot be used to measure distances across the curvature of the Earth without mathematical transformation that accounts for the bulging of the surface. The Earth Globe does exactly such mathematical transformation while maintaining one-to-one correspondence between the flat and spherical surfaces. That ensures that all towns, rivers, streets, topography, ..etc, are maintained, yet with different visualization of the data.

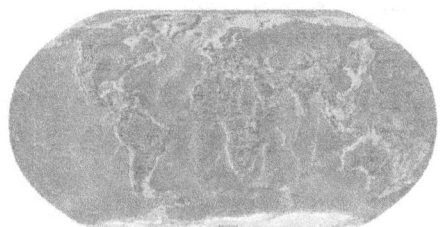

Riemann Surface

Euclidean Surface

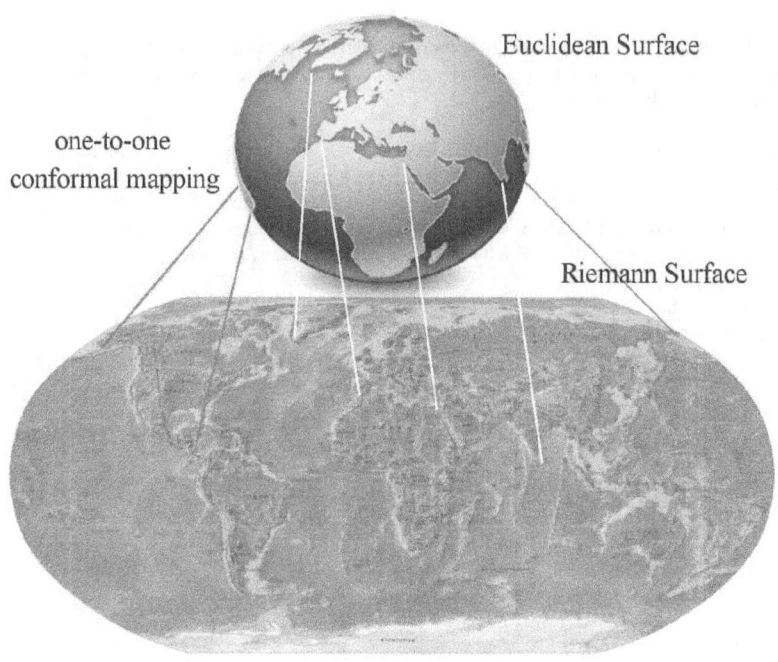

Conformal mapping is one-to-one transformation of points from one surface (Euclidean plane, in this example) to another surface (Riemann Sphere, in this example). It is a sort of encryption of data, which must also be decrypted through transformation and reverse transformation, with the benefit of gaining different visualization of the data that could be very convenient for expedient applications.

CHAPTER 3

ALGEBRAIC OPERATIONS ON COMPLEX NUMBERS

3.1. Addition of complex numbers

Let two complex numbers

$z_1 = x_1 + i y_1$ and
$z_2 = x_2 + i y_2$

Adding gives

$z_1 + z_2 = (x_1 + x_2) + i (y_1 + y_2)$

$z_1 + z_2 = (\text{sum of real}) + i (\text{sum of imaginary})$

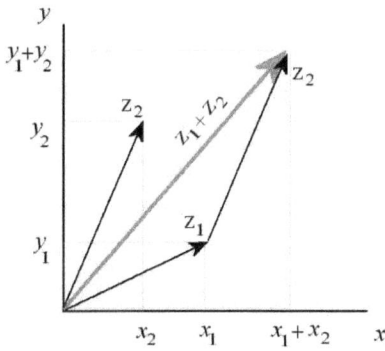

Thus, it is no surprise that the addition of complex numbers follows the law of **vector addition.**

In the above triangle between the complex numbers $z_1 + z_2$, z_1, and z_2, one could quickly discern the fact that the modulus of $z_1 + z_2$ must always be less than the sum of the modulus of z_1 and the modulus of z_2

Thus, $| z_1 + z_2 | \leq | z_1 | + | z_2 |$

Also, since $z_1 = (z_1 + z_2) + (- z_2)$

Then,

15

$| z_1 | \le | z_1 + z_2 | + | z_2 |$
Therefore,

$| z_1 + z_2 | \ge | z_1 | - | z_1 |$
Also,

$| z_1 + z_2 | \ge | \, | z_1 | - | z_1 | \, |$

3.2. Reciprocal of a complex number

$$\text{Let } z = x + i\,y = r\,(\,cos\,\theta\, + i\,sin\,\theta\,)$$

Then,

$$\frac{1}{z} = \frac{1}{r(\cos\theta + i\sin\theta)} = \frac{1}{r}\cdot\frac{\cos\theta - i\sin\theta}{\cos^2\theta + \sin^2\theta}$$

We obtained the above result by multiplying both the numerator and denominator by the conjugate of (*cos θ* + *i sin θ*); that is (*cos θ* - *i sin θ*), which resulted in yielding a unity denominator, since

$$cos^2\,\theta\, + \, sin^2\theta\, = 1.$$

Then,

$$\frac{1}{z} = \frac{1}{r}.(\cos\theta - i\sin\theta) = \frac{1}{r}.[\cos(-\theta) + i\sin(-\theta)]$$

Therefore, the reciprocal of a complex number has the **modulus** $\left| \dfrac{1}{z} \right|$ and the **argument –θ,** such that

16

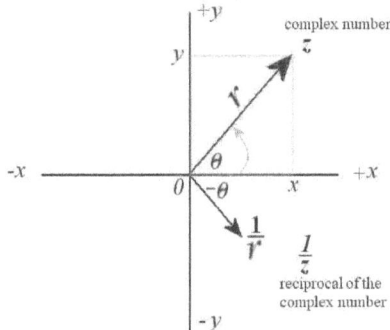

Note: the reciprocal of a complex number implies that the **modulus** reduced (if the original modulus was greater than unity) and the **angle** is reversed in sign.

3.3. The product of complex numbers

Let two complex variables be

$$z_1 = x_1 + i\, y_1$$
$$= r_1 \,(\, cos\,\theta_1 \; + i\, sin\,\theta_1\,)$$

and

$$z_2 = x_2 + i\, y_2$$
$$= r_2 \,(\, cos\,\theta_2 \; + i\, sin\,\theta_2\,)$$

Then, the Cartesian product of z_1 and z_2, gives:

$$z_1\, z_2 = (x_1\, x_2 - y_1\;\, y_2\,) + i\,(x_1\, y_2 + x_2\, y_1)$$

Whereas, the Polar product of z_1 and z_2, gives:

$$z_1\, z_2 = r_1\, r_2\,(\, cos\,\theta_1 \; + i\, sin\,\theta_1\,)\,(\, cos\,\theta_2 \; + i\, sin\,\theta_2\,)$$
$$= r_1\, r_2\,[\,(\, cos\,\theta_1\; cos\,\theta_2 - sin\,\theta_1\; i\, sin\,\theta_2\,)]\; + i\,[\, sin\,\theta_1\; cos\,\theta_2 + \; cos\,\theta_1\; sin\,\theta_2\,)]$$

Using the trigonometric rules of angle addition, we could reduce the above form to the following:

$$z_1\, z_2 = r_1\, r_2\, [\; cos\, (\, \theta_1\; + \theta_2\,)\; + i\, sin\, (\, \theta_1 + \theta_2\,)]$$

Therefore, the product of the two complex numbers z_1 and z_2 yields

$$\text{modulus} \mid z_1\, z_2 \mid\, = r_1\, r_2 = \mid z_1 \mid \; \mid z_2 \mid$$

$$\text{arg. } z_1\, z_2 = \theta_1\; + \theta_2 = \text{arg. } z_1 + \text{arg. } z_2$$

3.4. Division of complex numbers

We already proved the following two axioms:

(1) $\dfrac{1}{z} = \dfrac{1}{r}.(\cos\theta - i\sin\theta) = \dfrac{1}{r}.[\cos(-\theta) + i\sin(-\theta)]$

(2) $z_1\, z_2 = r_1\, r_2\, [\; cos\, (\, \theta_1\; + \theta_2\,)\; + i\, sin\, (\, \theta_1 + \theta_2\,)]$

Therefore, based on (1) and (2), we could immediately conclude that

$$\dfrac{z_1}{z_1} = \dfrac{r_1}{r_2}.[\cos(\theta_1 - \theta_2) + i\sin(\theta_1 - \theta_2)]$$

Thus, the division of two complex numbers z_1 and z_2 yields

$$\text{modulus} \mid \dfrac{z_1}{z_1} \mid\, = \dfrac{r_1}{r_2} = \dfrac{\mid z_1 \mid}{\mid z_1 \mid}$$

$$\text{arg. } z_1\, z_2 = \theta_1\; - \theta_2 = \text{arg. } z_1 - \text{arg. } z_2$$

3.5. The nth roots of complex numbers

$$z^{\frac{1}{n}} = r^{\frac{1}{n}}\, (\; cos\, \theta\; + i\, sin\, \theta\,)^{\frac{1}{n}}$$

$$= r^{\frac{1}{n}}\, [\, cos\, (\, \theta\; + 2s\pi\,) + i\, sin\, (\, \theta\; + 2s\pi\,)]^{\frac{1}{n}}, \text{ where } s = 0, 1, 2, 3, ..., n\text{-}1.$$

$$z^{\frac{1}{n}} = r^{\frac{1}{n}} [\cos \frac{\theta + 2s\pi}{n} + i \sin \frac{\theta + 2s\pi}{n}]$$

3.6. Conjugate complex numbers

If, $z = x + iy$, **the conjugate** $\bar{z}$ is given by $\bar{z} = x - iy$

Therefore,

$$|z| = |\bar{z}| = \sqrt{x^2 + y^2}$$

$$z + \bar{z} = 2x$$

$$z \bar{z} = x^2 + y^2$$

Thus, the **sum** and **product** of two **conjugate** complex numbers are **real** numbers.

3.7. Polynomials of complex numbers

Polynomials or rational integral functions of complex numbers **z** are given by

$$a_0 z^n + a_1 z^{n-1} + \ldots\ldots + a_{n-1} z + a_n$$

where, **n** is a +ve integer and $a_0, a_1, \ldots\ldots a_n$ are in general complex numbers.

3.8. Rational Functions of complex numbers

Rational functions take the form $\dfrac{P(z)}{Q(z)}$, where P(z) and Q(z) are polynomials in the complex number (z).

3.9. Exponential Functions of complex numbers

The exponential function of complex numbers is defined in the terms of the following polynomial:

$$e^z = 1 + z + \frac{z^2}{2!} + \frac{z^3}{3!} + \ldots..$$

Therefore, the product of two exponential functions corresponds to the additions of the exponents as follows:

19

$$e^{z_1} \cdot e^{z_2} = (1 + z_1 + \frac{z_1^2}{2!} + \frac{z_1^3}{3!} + ...) \ (1 + z_2 + \frac{z_2^2}{2!} + \frac{z_2^3}{3!} + ...)$$

$$= 1 + z_1 + z_2 + (\frac{z_1^2}{2!} + \frac{z_2^2}{2!} + z_1 z_2) + ...$$

$$= 1 + (z_1 + z_2) + \frac{1}{2!} \ (z_1^2 + z_2^2 + 2 \, z_1 z_2) + ...$$

$$= 1 + (z_1 + z_2) + \frac{1}{2!} \ (z_1 + z_2)^2 + ...$$

$$= e^{z_1 + z_2}$$

Therefore,

$$e^z = e^{x + iy} = e^x. \ e^{iy} = e^x. \ (cos \ y + i \ sin \ y)$$

Therefore,

modulus $| e^z | = e^x$

argument of $e^z = y$

Note:

Euler's formula is known as

$$e^{i\theta} = cos \ \theta + i \ sin \ \theta$$

Further,

$$e^{z + 2\pi i} = e^z . e^{2\pi i} = e^z \ (cos \ 2\pi + i \ sin \ 2\pi) = e^z$$

Hence, the exponential function $= e^z$ is periodic, with imaginary period equal to $2\pi i$

3.10. Hyperbolic Functions of complex numbers

The hyperbolic functions of complex variables are defined as:

$$cosh \ z = \frac{e^z + e^{-z}}{2}$$

$$sinh \ z = \frac{e^z - e^{-z}}{2}$$

$$tanh \ z = \frac{sinh \ z}{cosh \ z} = \frac{e^z - e^{-z}}{e^z + e^{-z}}$$

Useful formulas are given by:

$$\cosh^2 z - \sinh^2 z = 1$$

$$\cos^2 z + \sin^2 z = 1$$

The relationships between the hyperbolic functions and the trigonometric functions are superbly demonstrated by the exponential functions of complex numbers.

Example 4

Separate the real and imaginary parts of the complex function

$$\cosh (\alpha + i \beta)$$

Solution

Let $\cosh (\alpha + i \beta)$, then using the definition the hyperbolic functions, we get

$$\cosh (\alpha + i \beta) = \frac{e^{\alpha + i\beta} + e^{-\alpha - i\beta}}{2}$$

$$= \frac{1}{2}[e^{\alpha + i\beta} + e^{-\alpha - i\beta}]$$

$$= \frac{1}{2}[e^{\alpha} e^{i\beta} + e^{-\alpha} e^{-i\beta}]$$

$$= \frac{1}{2}\{e^{\alpha} (\cos \beta + i \sin \beta) + e^{-\alpha} [\cos(-\beta) + i \sin(-\beta)]\}$$

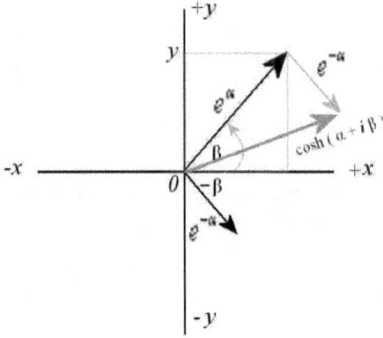

Representation of cosh (α + i β) in terms of the angles β and – β and the moduli e^{α} and $e^{-\alpha}$.

Let sin (α + i β), then be the definition the hyperbolic functions, we get

$$\sinh(\alpha + i\beta) = \frac{e^{\alpha+i\beta} - e^{-\alpha-i\beta}}{2}$$

$$= \frac{1}{2}[e^{\alpha+i\beta} - e^{-\alpha-i\beta}\,]$$

$$= \frac{1}{2}[e^{\alpha}e^{i\beta} - e^{-\alpha}e^{-i\beta}\,]$$

$$= \frac{1}{2}\{e^{\alpha}(\cos\beta + i\sin\beta) - e^{-\alpha}[\cos(-\beta) + i\sin(-\beta)]\}$$

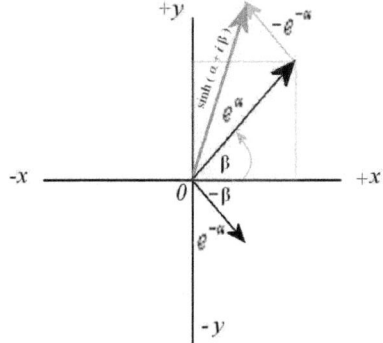

Representation of sinh ($\alpha + i\,\beta$) in terms of the angles β and $-\beta$ and the moduli e^{α} and $e^{-\alpha}$.

3.11. Trigonometric Functions of complex numbers

The main trigonometric functions sin and cos can be represented in terms of the exponential function of complex variables as follows:

$$\cos\ z = \frac{e^{iz} + e^{-iz}}{2}$$

$$i\sin\ z = \frac{e^{iz} - e^{-iz}}{2}$$

This can be proven directly from the polynomial formulas of the sin, cos, and exponential function e. Consequently,

$\cos\ z = \cosh\ (i\,z)$ and $\cos\ (iz) = \cosh z$

$i\sin\ z = \sinh\ (i\,z)$ and $\sin\ (iz) = i\sinh z$

Example 5

Prove that all zeroes of $\sin z$ are real numbers, then determine their values.

Solution

By a zero of $\sin z$ we mean a value of z which makes $\sin z$ equal to zero.
Therefore, from the above definition of $\sin z$ in terms of the exponential functions e, we get

23

$e^{iz} - e^{-iz} = 0$

i.e.,

$e^{iz} = e^{-iz}$

Therefore,

$e^{2iz} = 1$

i.e.,

$e^{2iz} = e^{2ik\pi}$

Therefore,

$2\,i\,z = 2\,i\,k\,\pi,$

Hence,

$z = k\,\pi = 0, \pm\,\pi, \pm 2\,\pi, \pm\,3\pi, \dots$

Example 6

Prove that

$$|\sin z\,|^2 = \sin^2 x + \sinh^2 y,$$

Where, $z = x + iy.$

Solution

From the properties of trigonometric functions, the sin of a sum can be decomposed to:

$\sin (x + iy) = \sin x \cos iy + \cos x \sin iy \dots\dots\dots\dots\dots(3.1)$

Then, using the above formula that link trigonometric functions with hyperbolic functions we get:

$\sin (x + iy) = \sin x \cosh y + i \cos x \sinh y$

From the above decomposed elements of z, the **modulus** $|\sin z|$ is given by

$|\sin z\,| = \sqrt{(\sin x \cosh y)^2 + (\cos x \sinh y)^2}$

24

$|\sin (x + iy)|^2 = \sin^2 x \cosh^2 y + \cos^2 x \sinh^2 y$ ....(3.2)

Substituting the following useful formulas in (3.2)

$\cosh^2 y = 1 + \sinh^2 y$

$\cos^2 x = 1 - \sin^2 x$

we get,

$|\sin (x + iy)|^2 = \sin^2 x (1 + \sinh^2 y) + (1 - \sin^2 x) \sinh^2 y$

$= \sin^2 x + \sin^2 x \sinh^2 y + \sinh^2 y - \sin^2 x \sinh^2 y$

$= \sin^2 x + \sinh^2 y$ ....which is the required proof of square of modulus

Also, the argument of sin z is given by equation (3.1) as:

$$arg.\ \sin z = \frac{complex - part - of - \sin z}{real - part - of - \sin z} = \frac{\cos x \sinh y}{\sin x \cosh y}$$

$arg.\ \sin z = \tanh y / \tan x$ ....(3.3)

No to confuse angles with trigonometric functions, but the angle, say θ, of sin z is given by

$\tan \theta - \tanh y / \tan x$

The above examples show the use of both **trigonometric functions** and **hyperbolic functions** through the implementation of theory of complex numbers.

| 3.12. Logarithmic Functions of complex numbers |

Let $w = u + iv = \ln z$ ....(3.4)

Therefore,

$z = e^{u + iv} = e^u e^{iv} = e^u (\cos v + i \sin v)$

We could return to the polar representation of z to get

$z = r (\cos v + i \sin v)$

This implies that

$r = e^u$ or $u = ln\ r$..........................(3.5)

$v = \theta + 2k\pi$

Hence,

$w = u + iv = ln\ z = ln\ r + i(\theta + 2k\pi)$

i.e., $ln\ z$ is many-valued function, with a **principal value** of

$ln\ z = ln\ r + i\theta$..........................(3.6)

Equation (3.6) represents **source /sink** of radiation emanating from $r = 0$, expanding circularly as r increases, each ray is directed at angle θ with respect to x-axis.

Example 7

Find the real and imaginary parts of the expression

$ln\ (1+ i).$

Solution

We will introduce the 45° as a mean of transforming the Cartesian form $(1+i)$ into polar form. Since, $\sin 45° = \cos 45° = \sin \dfrac{\pi}{4} = \dfrac{1}{\sqrt{2}}$, then we could write:

$(1+ i) = \sqrt{2}\ (\dfrac{1}{\sqrt{2}} + i\dfrac{1}{\sqrt{2}})$

$= \sqrt{2}\ (\cos \dfrac{\pi}{4} + i \sin \dfrac{\pi}{4})$

Using the formulas obtain above for, $z = r(\cos v + i \sin v)$, $ln\ z = ln\ r + i\ \theta$, we immediately conclude that

$ln\ (1+ i) = ln\sqrt{2} + i\ \dfrac{\pi}{4}$

$\qquad = \dfrac{1}{2}\ ln\ 2 + i\ \dfrac{\pi}{4}.$

| 3.13. Inverse Hyperbolic Functions of complex numbers |

It will be latter shown that the inverse hyperbolic functions of complex numbers can be defined as follows:

$$\sinh^{-1} z = \ln (z + \sqrt{z^2 + 1})$$
$$\cosh^{-1} z = \ln (z + \sqrt{z^2 - 1})$$
$$\tanh^{-1} z = \frac{1}{2} \ln \left(\frac{1+z}{1-z}\right)$$

3.14. Inverse Trigonometric Functions of complex numbers

Those are defined as: $\sin^{-1} z$, $\cos^{-1} z$, and $\tan^{-1} z$
Starting with: $\sin^{-1} z$, let

$$w = \sin^{-1} z$$

Therefore,

$$z = \sin w$$
$$= \frac{e^{iw} - e^{-iw}}{2i}$$

$$2 i z = e^{iw} - e^{-iw}$$

Multiplying both side by e^{iw}, we get

$$e^{2iw} - 2 i z e^{iw} - 1 = 0 \quad \dots\dots\dots\dots\dots\dots\dots (3.7)$$

Equation (3.7) is a quadratic equation, which have roots of the form:

$$e^{iw} = \frac{2iz \pm \sqrt{4i^2 z^2 + 4}}{2}$$
$$= iz \pm \sqrt{1 - z^2}$$

Therefore,

$$i w = \ln (iz \pm \sqrt{1 - z^2})$$

$$w = \sin^{-1} z = \frac{1}{i} \ln (iz \pm \sqrt{1 - z^2})$$

3.15. Complex Exponent Functions of complex numbers

27

The complex exponential function z^α can be represented in terms of the natural exponential form of e as follows.

$$z^\alpha = e^{\alpha \, ln \, z}$$

This can be proven by simply applying the logarithmic operator of both sides of the above equation and arranging the exponents.

3.16. Algebraic and Transcendental Functions of complex numbers

The difference between algebraic functions and transcendental function resides in the ability to separate variables independent from each other. For example, the function $w = \sqrt{z}$, which can be written as $w^2 - z = 0$, is clearly **algebraic** since w and z can be expressed in the terms of the other without being dependent on that other variable.

Any solution w for the polynomial equation

$$P_0(z) \, w^n + P_1(z) \, w^{n-1} + \ldots + P_{n-1}(z) \, w + P_n(z) = 0$$

Where, n is a positive integer and $P_0(z)$, $P_1(z)$, $\ldots$, $P_n(z)$, are polynomials of z, is an algebraic functions of z.

Other functions are **transcendental** or non-algebraic. e.g., $ln \, z$, $\sin z$, ...

Functioned derived from other functions by performing a finite number of operations such as addition, subtraction, multiplication, division, or extraction of roots are called **elementary** functions. e.g., $\sin z + 2 \, \cosh^{-1} z$.

3.17. Definition of Limit of complex functions

Let $f(z)$ be a single-valued function of z defined from all points in the immediate neighborhood of z_0 itself.

We say that $f(z) \rightarrow l$ as $z \rightarrow f(z)$ if given any positive number ε, however small, there exists $\delta = \delta(\varepsilon)$ that is dependent on ε, such that $| f(z) - l | < \varepsilon$ for all z satisfying $| z - z_0 | < \delta$, irrespective of the way in which z approaches its limit z_0.

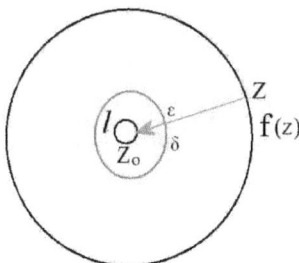

We write $\lim\limits_{z \to z_o} f(z) = l$

Example 8

$$\lim_{z \to 0} \frac{\sin x}{x} = 1$$

In this example, as sin x approaches x = 0, it gets smaller before it vanishes exactly at x = 0 , in the same manner that x also gets smaller before in vanishes exactly at x = 0. Therefore, the ration of the two infinitesimal values ε of the sine, and δ of the **x,** yields 1. This conclusion could also be proven algebraically by the use of differential calculus.

The limit of functions is a microscopic representation of the function near singularities where the numerator and denominators portions of the function race to the singularity leading to either undefined function of discontinuity, finite value, or vanishing function.

3.18. Continuity of complex functions

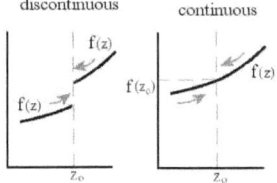

$f(z)$ is said to be continuous as $z = z_0$ if $\lim\limits_{z \to z_0} f(z) = f(z_0)$, both from the right and the left. This implies that f(z) should be defined at $z = z_0$.

A function f(z) is said to be **continuous** in a give **region** if it is continuous at all points in that region. This is generalization of the previous axiom of continuity at an **individual point**.

3.19. Definition of a domain of complex numbers

A domain is a surface of points D satisfying the following two properties:

(1) If P and Q are any points of a domain, then we can join P and Q by a curve, where all the points on that curve belong to the domain.

(2) If A is a point of the domain, then with A as a center, we can draw some circles such that all points in the interior of those circles belong to the domain

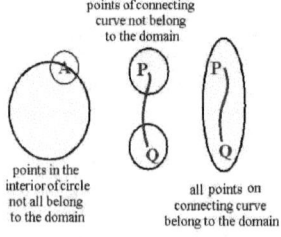

3.20. Definition of derivatives of complex functions

Derivatives of functions are the most powerful mathematical tools in physical sciences as they represent the instantaneous rate of change in functions, which amounts to the microscopic alteration in the function that could be integrated to reconstruct the macroscopic outcome of that function.

The derivative of f(z) with respect to z is defined as

$$\frac{\partial f(z)}{\partial z} = \lim_{\Delta z \to 0} \frac{f(z + \Delta z) - f(z)}{\Delta z}$$

This definition of derivative applies irrespective of the way in which $\Delta z - 0$. Suppose that Q approaches P parallel to the x-axis, then $\Delta z = \Delta x$. Then suppose that

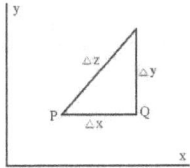

$f(z) = u(x, y) + i v(x, y)$ ....(3.8)

u and **v** are conjugate functions.

Apparently, since z depends on x and y, then $\dfrac{\partial f(z)}{\partial z}$ must have derivatives with respect to x and y, separately.

Since,

$z = x + i y$

Then,

$\Delta z = \Delta x + i \Delta y$

Now, keeping y unchanged implies that $\Delta y = 0$ and $\Delta z = \Delta x$. We get

$$\frac{\partial f(z)}{\partial x} = \lim_{\Delta x \to 0} \frac{u(x+\Delta x, y) + iv(x+\Delta x, y) - u(x, y) - iv(x, y)}{\Delta x}$$

$$\frac{\partial f(z)}{\partial x} = \frac{\partial u}{\partial x} + i \frac{\partial v}{\partial x}$$ ....(3.9)

Now, keeping x unchanged implies that $\Delta x = 0$ and $\Delta z = i \Delta y$

$$\frac{\partial f(z)}{\partial y} = \lim_{\Delta y \to 0} \frac{u(x, y+\Delta y) + iv(x, y+\Delta y) - u(x, y) - iv(x, y)}{i\Delta y}$$

$$\frac{\partial f(z)}{\partial y} = \frac{1}{i}(\frac{\partial u}{\partial y} + i \frac{\partial v}{\partial y})$$

Using the property: $i^2 = -1$

$$\frac{\partial f(z)}{\partial y} = \frac{\partial v}{\partial y} - i \frac{\partial u}{\partial y}$$ ....(3.10)

It should be noted that equations (3.9) and (3.10) were obtained on the assumption that the limit when Δz approaches zero does not depend on whether such approach was through the x-direction or the y-direction. Hence, we could equate equations (3.9) and (3.10) to get;

$$\frac{\partial v}{\partial y} - i\frac{\partial u}{\partial y} = \frac{\partial u}{\partial x} + i\frac{\partial v}{\partial x}$$

Equating the real and imaginary parts, we get

$$\frac{\partial v}{\partial y} = \frac{\partial u}{\partial x} \quad \text{and} \quad \frac{\partial u}{\partial y} = -\frac{\partial v}{\partial x} \quad ....(3.11)$$

Those are known as **Cauchy-Riemann** Equations, which must be satisfied in order that the complex function; $f(z) = u(x, y) + i v(x, y)$, to be complex differentiable.

3.20.1. Cauchy-Riemann Equations

The Cauchy-Riemann Equations (3.8) provides the conditions for differentiability of a function of complex variable. Further differentiation of the Cauchy-Riemann Equations yields conditions for continuity of the first derivatives on the points of reflection.

Thus, differentiating (3.11), with respect to x, we get:

$$\frac{\partial^2 v}{\partial y \partial x} = \frac{\partial^2 u}{\partial x^2} \quad \text{and} \quad \frac{\partial^2 u}{\partial y \partial x} = -\frac{\partial^2 v}{\partial x^2} \quad ....(3.12)$$

Thus, differentiating (3.11), with respect to y, we get

$$\frac{\partial^2 v}{\partial y^2} = \frac{\partial^2 u}{\partial x \partial y} \quad \text{and} \quad \frac{\partial^2 u}{\partial y^2} = -\frac{\partial^2 v}{\partial x \partial y} \quad ....(3.13)$$

3.20.2. Laplace Equation

Equations (3.12) and (3.13) immediately lead to **Laplace equations** by noticing that

$$\frac{\partial^2 v}{\partial x^2} + \frac{\partial^2 v}{\partial y^2} = 0, \text{ which can be formulated as } \nabla^2 v = 0........................(3.14)$$

$$\frac{\partial^2 u}{\partial x^2} + \frac{\partial^2 u}{\partial y^2} = 0, \text{ which can be formulated as } \nabla^2 u = 0 \dots\dots\dots\dots\dots\dots(3.15)$$

As we see, the function of complex variable, $f(z)$ in equation (3.8) yielded the two Laplace equations (3.14) and (3.15), which imply that both the real function u and the imaginary function v must satisfy the Laplace equation under the continuity conditions.

In fluid mechanics, Laplace equation is satisfied by $\nabla^2 q = 0$. But, in complex functions, $q = u + i\,v$, which implies two Laplace equations, one for the real function u, another for the complex function $\underline{v}$, such that $\nabla^2 u = 0$ and $\nabla^2 v = 0$.

We already denoted that u and v are **conjugate functions**. Then we proved that u and v each satisfies Laplace equations are thus called **harmonic functions**.

The above attained properties of conjugate functions will be proven crucial to many applications in **fluid mechanics, thermodynamics**, and **electromagnetics** where two perpendicular functions constitute the fundamental physical system of variable.

The utility of the function of complex variable, $f(z) = u(x, y) + i\,v(x, y)$, could be summed as follows:

(1) In fluid mechanics, the flow vector or fluid velocity is orthogonal on the lines of constant pressure, or isobaric surface. Thus, u represents the fluid flow rate and v the pressure distribution function.

(2) In electrodynamics, the flow of electric energy is orthogonal on the flow of magnetic energy. Thus, u represents the electric field and v the magnetic field.

(3) In thermodynamics, the flow of thermal energy is orthogonal on the surfaces of isotherms. Thus, u represents the temperature distribution in the medium and v the thermal flow along the temperature gradient. .

In order to prove, although that is obvious, that the curves of u and v are **orthogonal** to each other, or cut each other at right angles, we will first assume that

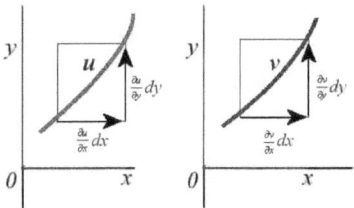

$u(x, y) = c$ ....(3.16)

Where, c is a constant, and u depends on both x and y.

Therefore,

$$du = \frac{\partial u}{\partial x}\, dx + \frac{\partial u}{\partial y}\, dy = 0,$$

Then

$$\frac{dy}{dx} = -\frac{\dfrac{\partial u}{\partial x}}{\dfrac{\partial u}{\partial y}},$$

From equation (3.11), we get

$$\frac{dy}{dx} = -\frac{\dfrac{\partial u}{\partial x}}{\dfrac{\partial u}{\partial y}} = \frac{\dfrac{\partial v}{\partial y}}{\dfrac{\partial v}{\partial x}} \quad ....(3.17)$$

Similarly, if we put

$v(x, y) = g$........................(3.18)

Where, g is a constant, and v depends on both x and y.

Therefore,

$$dv = \frac{\partial v}{\partial x}\, dx + \frac{\partial v}{\partial y}\, dy = 0,$$

Then

$$\frac{dy}{dx} = -\frac{\dfrac{\partial v}{\partial x}}{\dfrac{\partial v}{\partial y}},$$

Also, from equation (3.11), we get

$$\frac{dy}{dx} = -\frac{\dfrac{\partial v}{\partial x}}{\dfrac{\partial v}{\partial y}} = \frac{\dfrac{\partial u}{\partial y}}{\dfrac{\partial u}{\partial x}} \quad ....(3.19)$$

Multiply equations (3.17) by (3.19), we get

$$\left(\frac{dy}{dx}\right)_u \cdot \left(\frac{dy}{dx}\right)_v = \frac{\frac{\partial v}{\partial y}}{\frac{\partial v}{\partial x}} \times \frac{-\frac{\partial x}{\partial v}}{\frac{\partial v}{\partial y}} = -1 \dots\dots\dots\dots\dots\dots(3.20)$$

Therefore, *u* is orthogonal on *v*, because the product of the slopes of the curves of *u* and those of *v* = -1.

3.21. Regular and Analytic functions

Regular analytic functions are perfect examples in demonstrating a sanitized theoretical utility of the functions of complex variables.

A function $f(z)$ which is single-values and which has derivatives at all points of a given domain is called regular or analytic function in that domain. As we can see, this very idealized situation since in practice functions do not come the way we wish and the borders of the domains remain unsolved territories.

Example 9

Graph the following conformal mapping function in the w and z planes

$$w = z^2 \quad \dots\dots\dots\dots\dots\dots\dots(3.21)$$

where,

$$w = u + i\,v$$
$$z = x + i\,y$$

Solution

This is a prototype graphic demonstration of conjugate functions that has great scope of applications.

Therefore,

$$u + i\,v = (x + i\,y)^2 = x^2 - y^2 + 2\,i\,x\,y,$$

Separating the real and imaginary parts, we get the two sets of curves of the two conjugate functions as follows

$$u = x^2 - y^2 \dots\dots\dots\dots\dots\dots(3.22)$$
$$v = 2\,x\,y \quad \dots\dots\dots\dots\dots\dots(3.23)$$

$$\frac{\partial u}{\partial x} = 2\,x, \frac{\partial^2 u}{\partial x^2} = 2$$

$$\frac{\partial u}{\partial y} = -2\,y, \frac{\partial^2 u}{\partial y^2} \ -2$$

$$\frac{\partial v}{\partial x} = 2\,y, \frac{\partial^2 v}{\partial x^2} = 0$$

$$\frac{\partial v}{\partial y} = 2\,x, \frac{\partial^2 v}{\partial y^2} = 0$$

Therefore, the **Laplace equations** are satisfied as follows:

$$. \ \frac{\partial^2 u}{\partial x^2} + \frac{\partial^2 u}{\partial y^2} = -2 + 2 = 0$$

$$\frac{\partial^2 v}{\partial x^2} + \frac{\partial^2 v}{\partial y^2} = 0 + 0 = 0$$

Graphing the two sets of curves of equations (3.22) and (3.23), requires changing the values of the constants c and g, each value corresponds to a set of curves.

$$u = x^2 - y^2 = c \ \dots\dots\dots\dots\dots\dots(3.24)$$
$$v = 2\,x\,y = g \ \quad \dots\dots\dots\dots\dots\dots(3.25)$$

Differentiating the two sets of equations gives the slopes of the curves.

$$\frac{dy}{dx}\Big|_u = \frac{x}{y} \ \text{and} \ \frac{dy}{dx}\Big|_v = -\frac{y}{x} \ \dots\dots\dots\dots\dots\dots(3.26)$$

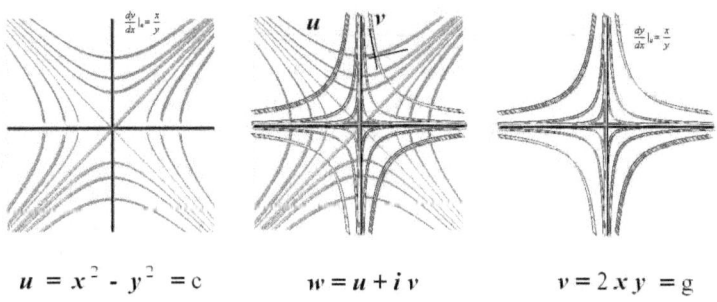

$$u = x^2 - y^2 = c \qquad\qquad w = u + i\,v \qquad\qquad v = 2\,x\,y = g$$

With the help of Equations (3.26), we get the equations of the tangents of **u** and **v** as follows.

(1) Equation of tangent to u: $\dfrac{y-y_1}{x-x_1} = \dfrac{x}{y}$

$(y - y_1)\, y = (x - x_1)\, x$

Therefore,

$x^2 - y^2 = x_1\, x - y_1\, y$

From equation (3.24), $x^2 - y^2 = c$, then

The tangent to u is given by: $x_1\, x - y_1\, y = c$

At the origin, $x_1 = y_1 = 0$, therefore, $c = 0$, hence, the slope becomes x_1/y_1 or 45°. This shows as the diagonal lines crossing the origin in the above left diagram. Another way of looking at the equation of the tangent of u, is to note that as $x_1\, x$ increases, $y_1\, y$ must increase, and vice versa.

(2) Equation of tangent to v: $\dfrac{y-y_1}{x-x_1} = -\dfrac{y}{x}$

$(y - y_1)\, x = -(x - x_1)\, y$

Therefore,

$2\, yx = x_1\, y + y_1\, x$

From equation (3.24), $2xy^2 = g$, then

The tangent to v is given by: $2\, yx = x_1\, y + y_1\, x = g$

In contrast to the tangent of the u, the tangent of v, as $x_1\, y$ increases, $y_1\, x$ must decrease, and vice versa. This is shown on the right diagram of the above plot.

Example 10

Graph the following conformal mapping function in the w and z planes

$$w = \dfrac{A}{z}, \quad \text{or} \quad w = \dfrac{A}{x+iy} = \dfrac{A(x-iy)}{x^2+y^2} \quad \dots\dots\dots\dots\dots\dots\dots(3.27)$$

Solution

Let us use our conventional representation of w and z. Hence,

$$w = \frac{A(x-iy)}{x^2+y^2} = u + i\,v.$$

Separating the real and imaginary parts, we get:

$u = \dfrac{Ax}{x^2+y^2}$, which gives $u\,(x^2+y^2) - A\,x = 0$.........................(3.28)

$v = -\dfrac{Ay}{x^2+y^2}$, which gives $v\,(x^2+y^2) + A\,y = 0$.........................(3.29)

Equation (3.28) represents a set of circles of various radii depending on the magnitude (A/u) centered on the x-axes at locations given by ($\pm$ A/ 2u, 0).

Similarly, Equation (3.29) represents a set of circles of various radii depending on the magnitude (A/v) centered on the y-axes at locations given by (0, $\pm$ A/ 2v).

This is represented in the following diagram.

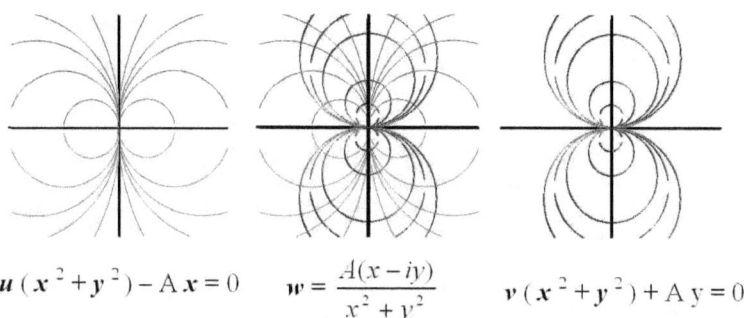

$u\,(x^2+y^2) - A\,x = 0$ $w = \dfrac{A(x-iy)}{x^2+y^2}$ $v\,(x^2+y^2) + A\,y = 0$

It clear that the creative formulation of the complex variable $w\,(z)$ results into very useful representations of conjugate functions which will be shown in may practical applications in aerodynamics, electrodynamics, and thermodynamics.

Example 11

Graph the following conformal mapping function in the w and z planes

$$w = \ln\frac{z-a}{z+a}, \quad \text{where "}a\text{" is real} \quad ....(3.30)$$

Solution

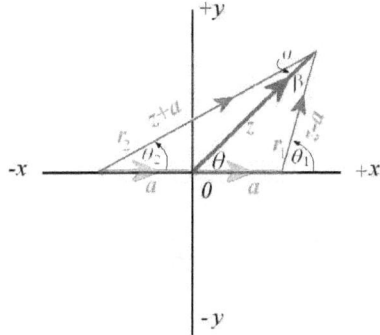

Equation (3.30) can be written as

$$w = \ln(z-a) - \ln(z+a) \dots\dots\dots\dots\dots\dots\dots(3.31)$$

Then we could the previous proven property of

$$\ln z = \ln r + i\theta$$

where,

$$z = r(\cos v + i \sin v),$$
$$r = e^u \quad \text{or} \quad u = \ln r,$$
and

$$v = \theta + 2k\pi$$

Therefore, using the notations in the above figure, we could write the two terms of equation (24) as follows:

$$\ln(z-a) = \ln(r_1) + i\theta_1$$
$$\ln(z+a) = \ln(r_2) + i\theta_2$$

Substituting by the above two terms in equation (3.31), we get

$$w = u + iv = \ln(r_1) + i\theta_1 - \ln(r_2) - i\theta_2 \dots\dots\dots\dots\dots\dots(3.32)$$

Separating the real and imaginary parts of equation (3.32), we get

39

$$u = ln \frac{r_1}{r_2} \quad \dots\dots\dots\dots\dots\dots\dots.(3.33)$$

$$v = \theta_1 - \theta_2 \dots\dots\dots\dots\dots\dots\dots.(3.34)$$

In order to plot the graphs of equations (3.33) and (3.34) we will rewrite the two equations in the Cartesian forms such that equation (3.33) becomes:

$$u = ln \frac{r_1}{r_2} = ln \frac{\sqrt{y^2 + (x-a)^2}}{\sqrt{y^2 + (x+a)^2}} = c \dots\dots\dots\dots\dots\dots.(3.35)$$

where c is a constant.

Also, equation (3.34) can be written in terms of the constant ratios of $r_1 : r_2 : 2\,a$, which is derived from the rules of sines of a triangle. This gives

$$\sqrt{y^2 + (x-a)^2} : \sqrt{y^2 + (x+a)^2} : 2\,a = g \dots\dots\dots\dots\dots\dots.(3.36)$$

Where, g is a constant.

Equation (3.35) corresponds to the Apollonius circles, left diagram in the figure below.

Equation (3.36) corresponds to circles centered at the ($\pm\,a$, 0).

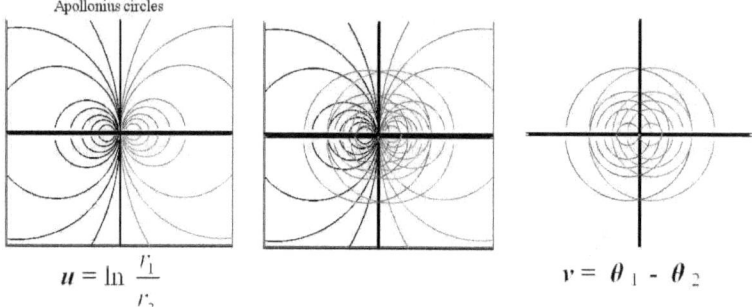

Apollonius circles

$$u = \ln \frac{r_1}{r_2} \qquad\qquad v = \theta_1 - \theta_2$$

3.22. Example on Laplace Equations

Example 12

Given that
$$z = x + iy$$
and

$$f(z) = u + iv$$

Use Laplace equations to prove that

$$u = x^4 - 6x^2 y^2 + y^4 \dots\dots\dots\dots\dots\dots(3.37)$$

is a possible differentiable function for **u**. Then find **v**.

Solution

Part I: <u>Satisfying Laplace's equation</u>

The four derivatives of equation (3.37) required to satisfy Laplace equations are obtained as follows:

$$\frac{\partial u}{\partial x} = 4x^3 - 12x\,y^2 \quad \dots\dots\dots\dots\dots\dots(3.38)$$

$$\frac{\partial u}{\partial y} = -12x^2\,y + 4y^3 \dots\dots\dots\dots\dots\dots(3.39)$$

$$\frac{\partial^2 u}{\partial x^2} = 12x^2 - 12\,y^2 \quad \dots\dots\dots\dots\dots\dots(3.40)$$

$$\frac{\partial^2 u}{\partial y^2} = -12x^2 + 12y^2 \quad \dots\dots\dots\dots\dots\dots(3.41)$$

Since, equations (3.40) and (3.41) yield a vanishing sum, therefore, equation (3.37) satisfies Laplace's equation, as follows:

$$\frac{\partial^2 u}{\partial x^2} + \frac{\partial^2 u}{\partial y^2} = 12x^2 - 12\,y^2 - 12x^2 + 12y^2 = 0\dots\dots\dots\dots\dots\dots(3.42)$$

Hence, **u**, equation (3.34), can be the real part of $f(z)$.

Part II: <u>Finding the imaginary part **v** of $f(z)$</u>

Applying Cauchy-Riemann Equations on equations (3.38) and (3.39), we get

$$\frac{\partial u}{\partial x} = 4x^3 - 12x\,y^2 = \frac{\partial v}{\partial y} \quad \dots\dots\dots\dots\dots\dots(3.43)$$

$$\frac{\partial u}{\partial y} = -12\,x^{2}\,y + 4\,y^{3} = -\frac{\partial v}{\partial x} \quad\text{................................(3.44)}$$

Integrating equation (3.44) with respect to x, we get

$$v = 4\,x^{3}\,y - 4\,x\,y^{3} + \varphi\,(y) \quad\text{........................(3.45)}$$

comparing the y-differentiated form of equation (3.43) with (3.45), we conclude that

$$\frac{\partial\varphi(y)}{\partial y} = 0,$$

Therefore,

$$v = 4\,x^{3}\,\text{y} - 4\,x\,y^{3}$$

Hence,

$$f(z) = u + i\,v$$
$$= x^{4} - 6\,x^{2}\,y^{2} + y^{4} + i\,(\,4\,x^{3}\,\text{y} - 4\,x\,y^{3}\,)$$

It should be noted that the expansion of $(x + i\,y)^{4}$ contains terms similar to the above polynomial as follows.

$$(x + i\,y)^{4} = x^{4} + 4\,x^{3}\,(\,i\,y\,) + \frac{4(4-1)}{2!}\,x^{2}\,(\,i\,y\,)^{2} + \frac{4(4-1)(4-2)}{3!}\,x\,(\,i\,y\,)^{3}$$

$$+ \ \frac{4(4-1)(4-2)(4-3)}{4!}\,(\,i\,y\,)^{4}$$

$$= x^{4} - 6\,x^{2}\,y^{2} + y^{4} + i\,(\,4\,x^{3}\,\text{y} - 4\,x\,y^{3}\,)$$

Therefore,

$$f(z) = u + i\,v = (\,x + i\,\text{y}\,)^{4}$$

CHAPTER 4

CONFORMAL MAPPING

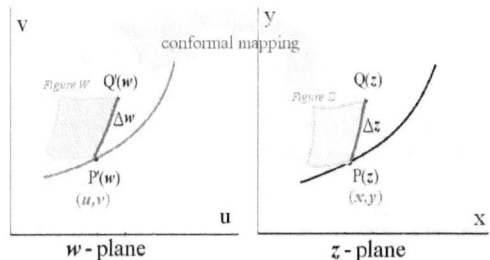

The mapping of functions between the two planes $z(x, y)$ and $w(u, v)$ is said to be conformal when:

(1) The angles of intersection of curves in the z-plane are equal to the corresponding angles of intersection of curves in the w-plane.

(2) Infinitesimal elements of area and shapes in the z-plane transform into similar infinitesimal elements of the same shape in the w-plane, such that

$$\frac{area\_of\_element\_in\_the\_w-plane}{area\_of\_corresponding\_element\_in\_the\_z-plane} = \left(\frac{dw}{dz}\right)^2 \quad \dots\dots\dots\dots\dots(4.1)$$

The above two conditions for conformal mapping are represented algebraically as follows.

Let
$w = f(z)$ be a regular function in a give domain.
$z = x + i y$
$w = u + i v$

Therefore,

$$u + i v = f(x + i y)$$

For this mapping equation to be conformal, every point P(z) in the z-plane, that is determined by the coordinates (x, y), corresponds to a point P^v(w) in the w-plane, determined by the coordinates (u, v).

When P(z) traces a curve in the z-plane, P`(w) traces a corresponding curve in the w-plane. The utility of such mapping lies in the ability to reduce complex shapes of surfaces into much simpler ones where mathematical analysis is more feasible, then, transform the performed analysis back to the complex form via the transform function.

To illustrate on a practical example, the airfoil of an aircraft could be simplified into a circular form by conformal mapping where the pressure and velocities studies could be done with great ease, than doing the same of a convoluted contour of airfoil. The circular function can be then mapped back to produce the real function of the airfoil ready for manufacturing purposes.

As P(z) moves a distance Δz to Q(z), on the z-plane,
P` (w) moves a distance $\Delta w$ to Q` (w), on the w-plane, such that

$$\Delta w = \frac{dw}{dz} \Delta z ....(4.2)$$

Where, $\frac{dw}{dz} = \frac{df(z)}{dz}$
Therefore,

$$|\Delta w| = |\frac{dw}{dz}| . |\Delta z| \qquad \text{and} ....(4.3)$$

$$\text{arg. } \Delta w = \textbf{\textit{arg. }} \frac{dw}{dz} . \textbf{\textit{arg. }} \Delta z. ....(4.4)$$

Thus, in the w-plane, the modulus of Δw is obtained by multiplying the distance PQ or Δz by the transform $|\frac{dw}{dz}|$. The argument of Δw is obtained adding the two arguments of $\frac{dw}{dz}$ and Δz.

This is true for infinitesimal vectors ensuing from P. Those are magnified or reduced, when mapped onto the w-plane, by the ratio $|\frac{dw}{dz}|$: 1. Those vectors also rotate through the same angle of $|\frac{dw}{dz}|$, or, arg. $|\frac{dw}{dz}|$.

Equations (4.1), (4.3), and (4.4) describe a simple process of **magnification** and **distortion** of images through the algebraic functional mapping demonstrated in the figure below. Equation (4.1) calculates the areal magnification (or reduction) ratio of the image upon the action of the differential operator $\frac{dw}{dz}$. Equation (4.3) calculates the linear magnification (or reduction) of lengths on the processed image through the same differential operator. Equation (4.4) calculates the angular changes in the image's curves due to the action of the same differential operator.

Confromal Mapping

$w(u,v)$ $z(x,y)$

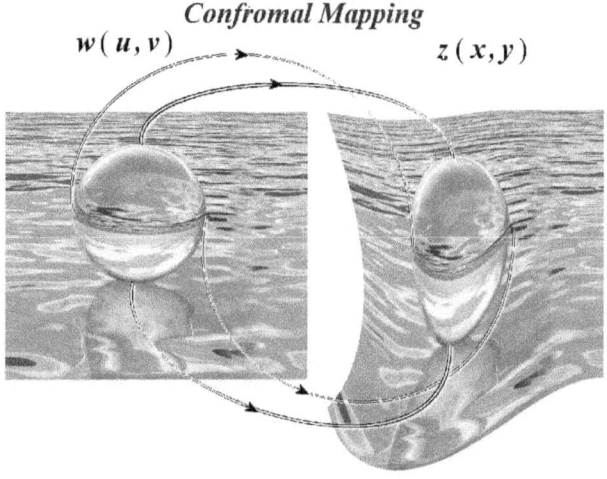

4.1. Jacobian or rates of change of **x, y**, on u, v

Now, we are settled on the two main properties of conformal mapping, which are **magnification** and **distortion**, which could also be given more meaningful mathematical names such as **factorization** and **deformation**, of something of that sort. We will then embark on the methods of quantifying those two properties such that we could calculate the amount of factorization and deformation of images operated upon by differential operators on the functions of complex variables.

Let $z = \varphi(w)$ and $w = u + iv$ ....(4.5)

i.e., $x + iy = \varphi(u + iv)$ ....(4.6)

Differentiating equation (4.6) with respect to u and v, consecutively, gives:

$$\frac{\partial x}{\partial u} + i\frac{\partial y}{\partial u} = \frac{\partial \varphi(w)}{\partial w} \cdot \frac{\partial w}{\partial u} \quad ....(4.7)$$

$$\frac{\partial x}{\partial v} + i\frac{\partial y}{\partial v} = \frac{\partial \varphi(w)}{\partial w} \cdot \frac{\partial w}{\partial v} \quad ....(4.8)$$

Differentiating equation (4.5) with respect to u and v, consecutively, gives:

$$\frac{\partial w}{\partial u} = 1 \quad \text{and} \quad \frac{\partial w}{\partial v} = i \quad ....(4.9)$$

45

From equations (4.7), (4.8), and (4.9), we get the Cauchy-Riemann Equations in reciprocal forms.

$$\frac{\partial x}{\partial u} = \frac{\partial y}{\partial v} \text{ and } \frac{\partial x}{\partial v} = -\frac{\partial y}{\partial u} \quad \ldots\ldots\ldots\ldots\ldots\ldots\ldots(4.10)$$

From equation (4.5),

$$|\frac{dz}{dw}|^2 = |\frac{\partial x}{\partial u}|^2 + |\frac{\partial y}{\partial u}|^2 \ldots\ldots\ldots\ldots\ldots\ldots\ldots(4.11)$$

Equations (4.10) and (4.11) give

$$|\frac{dz}{dw}|^2 = |\frac{\partial x}{\partial u}|^2 + |\frac{\partial y}{\partial u}|^2 = \frac{\partial x}{\partial u}\cdot\frac{\partial x}{\partial u} + \frac{\partial y}{\partial u}\cdot\frac{\partial y}{\partial u}$$

$$|\frac{dz}{dw}|^2 = \frac{\partial x}{\partial u}\cdot\frac{\partial y}{\partial v} - \frac{\partial y}{\partial u}\cdot\frac{\partial x}{\partial v}$$

$$|\frac{dz}{dw}|^2 = \begin{vmatrix} \frac{\partial x}{\partial u} & \frac{\partial y}{\partial u} \\ \frac{\partial x}{\partial v} & \frac{\partial y}{\partial v} \end{vmatrix} = \frac{\partial(x,y)}{\partial(u,v)} \quad \ldots\ldots\ldots\ldots\ldots\ldots\ldots(4.12)$$

The above determinant is known as *Jacobian* of *x* and *y* with respect to *u* and *v*. The Jacobian in equation (4.12) gives the area of the differential element in the *z*-plane.

Jacobian of *x* and *y* with respect to *u* and *v*.

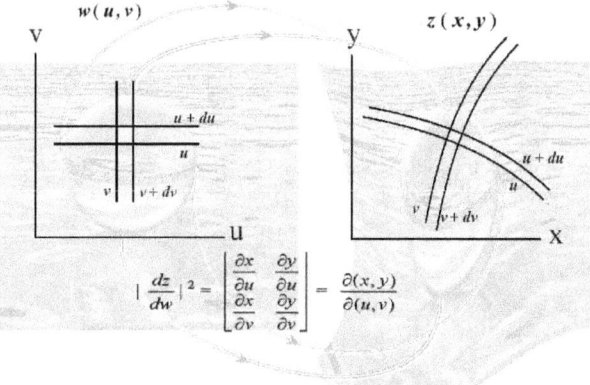

$$| \frac{dz}{dw} |^2 = \begin{vmatrix} \frac{\partial x}{\partial u} & \frac{\partial y}{\partial u} \\ \frac{\partial x}{\partial v} & \frac{\partial y}{\partial v} \end{vmatrix} = \frac{\partial(x,y)}{\partial(u,v)}$$

4.2. **Example** of mapping complex functions $z(x, y)$ on a quadratic function $w(u, v)$

Example 13

If $z = w^2$, prove that the curves of constant *u* and the curves of constant *v* are conicl parabolas and calculate the area between the two parabolas $u = 3$ and $v = 2$.

Solution

Let $x + iy = (u + iv)^2 = u^2 - v^2 + 2iuv$

$$x = u^2 - v^2$$
$$y = 2uv$$

Eliminating *v* between the expressions of *x* and *y* in the <u>case of constant *u*</u>, we get

$v = y / 2u$, which upon substituting in the expression of *x*, gives
$x = u^2 - y^2 / 4u^2$

Arranging the terms in the above expression, we get

$$y^2 = -4u^2(x - u^2) \dots\dots\dots\dots\dots\dots(4.13)$$

Equation (4.13) represents a set of parabolas with vertices located at $(u^2, 0)$, symmetrical around the x-axes, the two arms of the parabolas diverge in the negative half of the x-axis.

Eliminating *u* between the expressions of *x* and *y* in the <u>case of constant *v*</u>, we get

47

$u = y / 2\,v$, which upon substituting in the expression of x, gives
$x = y^2 / 4\,v^2 - v^2$

Therefore,

$$y^2 = 4\,v^2\,(x + v^2)\ \dots\dots\dots\dots\dots\dots(4.14)$$

Equation (4.14) represents a set of parabolas with vertices located at $(-v^2, 0)$, symmetrical around the x-axes, the two arms of the parabolas diverge in the positive half of the x-axis.

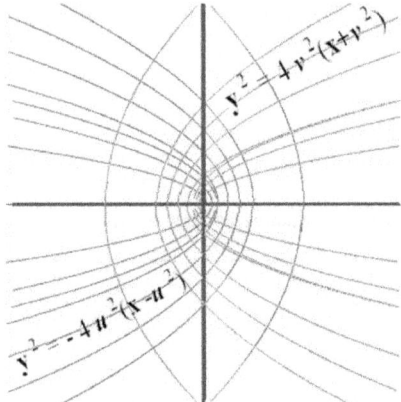

      The area between the two curves of equations (4.13) and (4.14) is calculated through the Jacobian as follows:

$$\left| \frac{dz}{dw} \right|^2 = \begin{vmatrix} \dfrac{\partial x}{\partial u} & \dfrac{\partial y}{\partial u} \\ \dfrac{\partial x}{\partial v} & \dfrac{\partial y}{\partial v} \end{vmatrix} = \frac{\partial(x, y)}{\partial(u, v)}$$

Where the partial derivates of x and y are obtained by differentiating the two initial equations:

$$x = u^2 - v^2$$
$$y = 2\,u\,v$$

Therefore,

$$\frac{\partial x}{\partial u} = 2\,u \qquad \frac{\partial x}{\partial v} = -2\,v \qquad \frac{\partial y}{\partial v} = 2\,u \qquad \frac{\partial y}{\partial u} = 2\,v$$

$$\left| \frac{dz}{dw} \right|^2 = \begin{vmatrix} 2u & 2v \\ -2v & 2u \end{vmatrix} = 4u^2 + 4v^2 \dots\dots\dots\dots\dots\dots(4.15)$$

Equation (4.15) gives the element of area under the two curves (4.13) and (4.14). Integrating equation (4.15) over the ranges $u = 0$ to 3 and $v = 0$ to 2, we get the area shared between those two specific curves.

The shared area between the 2 parabolas
$$= 2 \int_{u=0}^{3} \int_{v=0}^{2} 4(u^2 + v^2)\,du\,dv$$

$$= 8 \int_{u=0}^{3} (u^2 v + \frac{1}{3}v^3)\,|_0^2 \; du$$

$$= 8 \int_{u=0}^{3} (2u^2 + \frac{8}{3})\,du$$

$$= 8 \left(\frac{2}{3}u^3 + \frac{8}{3}u \right) |_0^3$$

$$= 8(18+8) = 208 \text{ units of area}$$

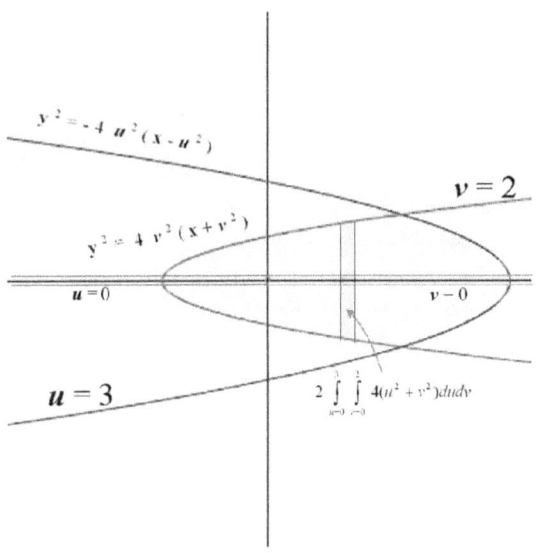

4.3. The Transformation of $w\,(u, v) = \exp(z\,(x, y))$

49

The transformation $w = e^z$ on the infinite strip in the z-plane is bounded by the two lines $y = 0$ and $y = 2\pi$.

$$w = e^z = e^{x + iy} = e^x\, e^{iy}$$
$$= e^x\,(\cos y + i \sin y)$$

Separating the real and imaginary parts, we get

$$u = e^x \cos y \dots\dots\dots\dots\dots\dots(4.16)$$
$$v = e^x \sin y \ \dots\dots\dots\dots\dots\dots(4.17)$$

Equations (3.16) and (3.17) give

$|w| = e^x$ argument. $w = y$

In order to eliminate y or x, we will square equations (4.16) and (4.17) and use the property of sines and cosines; $\sin^2 y + \cos^2 y = 1$. Therefore,

$$u^2 = e^{2x} \cos^2 y$$

$$v^2 = e^{2x} \sin^2 y$$

Adding the two equations, we get

$$u^2 + v^2 = e^{2x} \dots\dots\dots\dots\dots\dots(4.18)$$

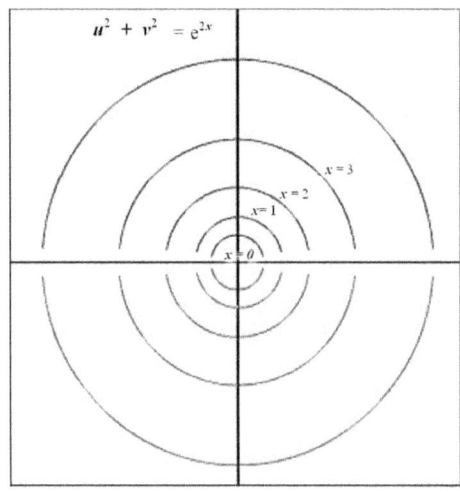

w-plane or the *v*-*u* plot of equation (4.18)

The mapping of $w = e^z$, at y = constant, transforms into circles in the *w*-plane with radii equivalent to e^x.

Taking the logarithm of both sides of (4.18), we get

$$x = \frac{1}{2} \ln (u^2 + v^2) \dots\dots\dots\dots\dots\dots(4.19)$$

Diving equation (4.17) by (4.16), we get

$$\tan y = \frac{v}{u} \qquad \dots\dots\dots\dots\dots\dots(4.20)$$

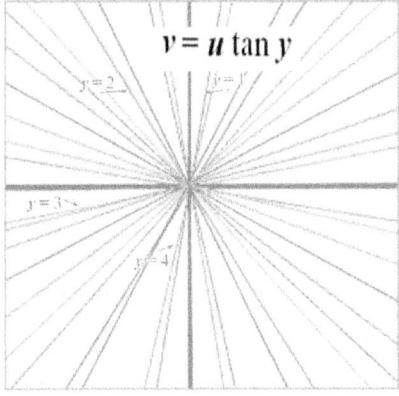

w-plane or the *v*-*u* plot of equation (20)

The mapping of $w = e^z$, at x = constant, transforms into rays emitting from the origin of the *w*-plane with angles varying as *y* changes.

Equation (4.18) gives two expressions for *u* and *v* in terms of x as follows

$$u = \sqrt{e^{2x} - v^2} \text{ and } v = \sqrt{e^{2x} - u^2} \dots\dots\dots\dots\dots\dots(4.21)$$

Equations (4.21) are substituted in equation (4.20) to give

$$y = \tan^{-1} \frac{v}{\sqrt{e^{2x} - v^2}} \quad \text{for constant } v \quad \dots\dots\dots\dots\dots(4.22)$$

$$y = \tan^{-1} \frac{\sqrt{e^{2x} - u^2}}{u} \quad \text{for constant } u \quad \dots\dots\dots\dots\dots(4.23)$$

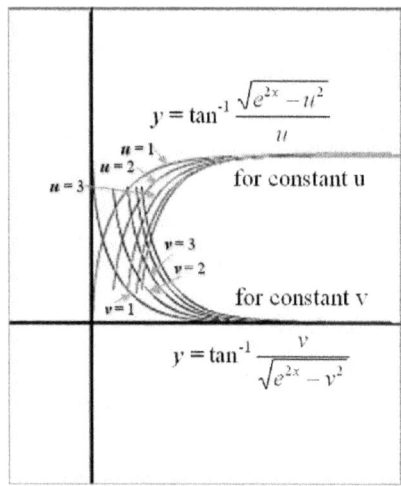

z-plane or the *y-x* plot of equations (4.22) and (4.23).

The Jacobian of e^z

Differentiating equation (4.18) with respect to *u* gives

$$2\, u = 2\, e^{2x} \frac{\partial x}{\partial u}$$

Or

$$\frac{\partial x}{\partial u} = \frac{u}{u^2 + v^2} \quad \dots\dots\dots\dots\dots(4.24)$$

Similarly, differentiating equation (4.18) with respect to *v* gives

$$\frac{\partial x}{\partial v} = \frac{v}{u^2 + v^2} \quad \dots\dots\dots\dots\dots(4.25)$$

Differentiating equation (4.20) with respect to *u* gives

$$\sec^2 y \frac{\partial y}{\partial u} = \frac{-v}{u^2} \quad \dots \dots \dots \dots \dots (4.26)$$

where, $\sec^2 y$ is obtained from the trigonometric relationship, shown in the following sketch.

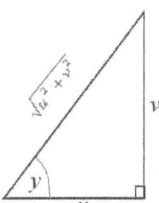

$$\sec^2 y = \frac{u^2 + v^2}{u^2} \quad \dots \dots \dots \dots \dots (4.27)$$

Thus, equations (4.26) and (4.27) give

$$\frac{\partial y}{\partial u} = \frac{-v}{u^2} \cdot \frac{u^2}{u^2 + v^2} = \frac{-v}{u^2 + v^2} \quad \dots \dots \dots \dots \dots (4.28)$$

Differentiating equation (4.20) with respect to v gives

$$\sec^2 y \frac{\partial y}{\partial v} = \frac{1}{u} \quad \dots \dots \dots \dots \dots (4.29)$$

Equations (4.27) and (4.29) give

$$\frac{\partial y}{\partial v} = \frac{1}{u} \cdot \frac{u^2}{u^2 + v^2} = = \frac{u}{u^2 + v^2} \quad \dots \dots \dots \dots \dots (4.30)$$

The Jacobian of e^z is obtained by substituting from equations (4.24), (4.25), (4.28), and (4.30), by the derivatives of x and y with respect to u and v in the Jacobian matrix as follows.

$$\left| \frac{dz}{dw} \right|^2 = \begin{vmatrix} \dfrac{\partial x}{\partial u} & \dfrac{\partial y}{\partial u} \\ \dfrac{\partial x}{\partial v} & \dfrac{\partial y}{\partial v} \end{vmatrix} = \begin{vmatrix} \dfrac{u}{u^2 + v^2} & \dfrac{-v}{u^2 + v^2} \\ \dfrac{v}{u^2 + v^2} & \dfrac{u}{u^2 + v^2} \end{vmatrix}$$

$$= \left(\frac{u}{u^2 + v^2} \right)^2 + \left(\frac{v}{u^2 + v^2} \right)^2$$

$$\left|\frac{dz}{dw}\right|^2 = \frac{u^2 + v^2}{(u^2 + v^2)^2} = \frac{1}{u^2 + v^2} \quad ....(4.31)$$

Thus, the area under the complex function e^z is obtained by integrating equation (4.31) as follows

$$= \int_{u=c1}^{u=c2} \int_{v=g1}^{v=g2} \frac{1}{u^2 + v^2} \, du \cdot dv$$

Thus, the conformal mapping of the **exponential functions** of equations (4.22) and (4.23) yielded the more manageable shapes of **circles** and radiations in figures (4.18) and (4.20). This feature of conformal mapping with be illustrated in details in applications on fluid mechanics.

4.4. The Transformation of $w\,(u, v) = \cosh\,(z\,(x, y))$

$w = u + i\,v = \cosh(x + i\,y)$
  $= \cosh x \cosh i\,y + \sinh x \sinh i\,y.........................(4.32)$

Substituting in equation (4.32), by the relationships

$\cos z = \cosh(i\,z)$ and $i \sin z = \sinh(i\,z)$

we get

$w = \cosh x \cos y + i \sinh x \sin y$

Therefore,

$u = \cosh x \cos y.........................(4.33)$
$v = \sinh x \sin y.........................(4.34)$

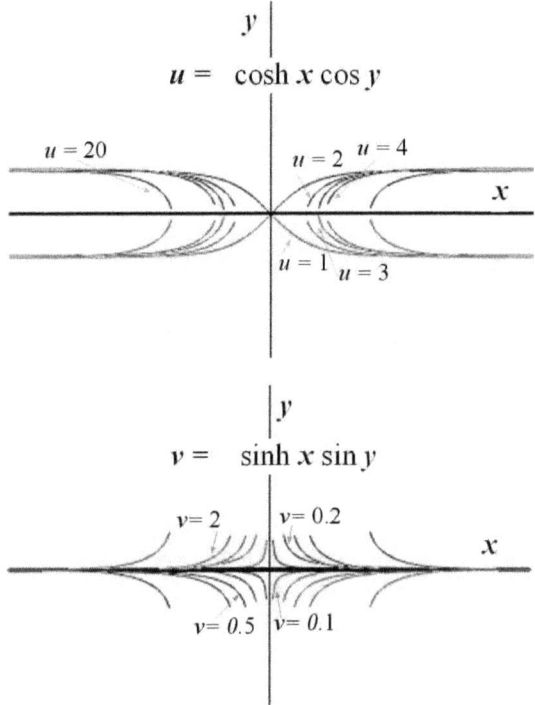

Eliminating y from equations (4.33) and (4.34), by squaring and adding and using the property

$\sin^2 y + \cos^2 y = 1$,

we get

$$\frac{u^2}{\cosh^2 x} + \frac{v^2}{\sinh^2 x} = 1 \dots\dots\dots\dots\dots(4.35)$$

Equation (4.35) represent a set of conic ellipses corresponding to x = constant, varying as the constant changes, with foci located at (± 1, 0).

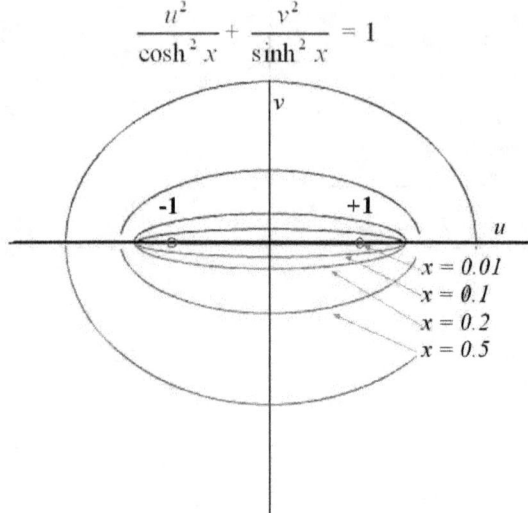

$$\frac{u^2}{\cosh^2 x} + \frac{v^2}{\sinh^2 x} = 1$$

Eliminating x from equations (4.33) and (4.34), by squaring and subtracting and using the property

$\cosh^2 x - \sinh^2 x = 1$, we get

$$\frac{u^2}{\cos^2 y} - \frac{v^2}{\sin^2 y} = 1 \ldots\ldots\ldots\ldots\ldots\ldots(4.36)$$

Equation (4.36) represent a set of conic hyperbolas corresponding to y = constant, varying as the constant changes, with foci located at (± 1, 0).

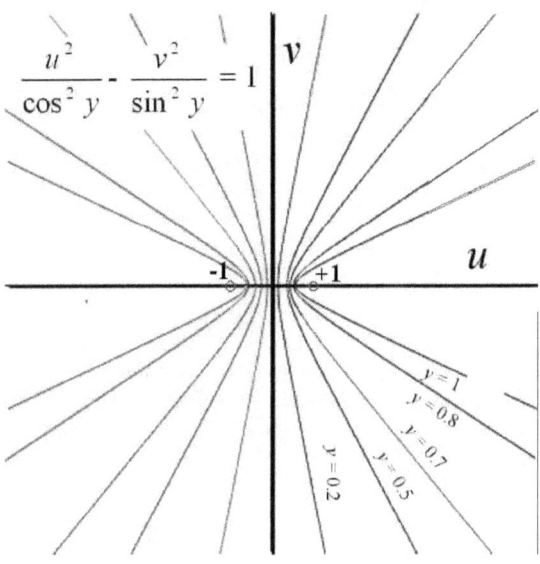

$$\frac{u^2}{\cos^2 y} - \frac{v^2}{\sin^2 y} = 1$$

v

u

-1   +1

$y = 1$
$y = 0.8$
$y = 0.7$
$y = 0.5$
$y = 0.2$

57

CHAPTER 5

INVERSE TRANSFORMATION

A **circle of inversion** of center O and radius k is used to invert the curve C onto the curve C', where point P located on C is mapped onto point P' on C' such that:

(1)   The three point O, P, and P' lay on a straight line and
(2)   The product of lengths of lines OP and OP' satisfies the relation:

$$(OP).(OP') = k^2.$$

When P traces the curve C, P' traces the inverse curve C'. The nature of the inverse curve depends on the origin O of inversion (the origin of k), which affects the position and sides of the inversed curve.

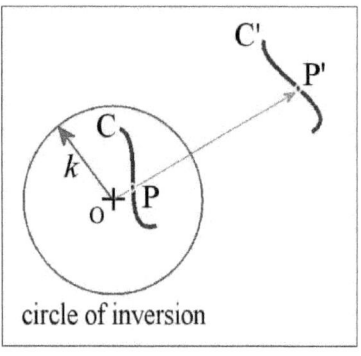

circle of inversion

Theorem I:

The inverse of a straight line with respect to a point on the **same line** is the same straight line.

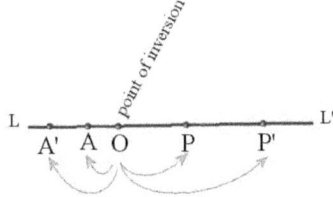

Proof

In the above figure, consider the LL' and an arbitrary point O chosen as an inversion point.

For any arbitrary point P on the line LL' with inversion point P', such that

$$(OP).(OP') = k^2$$

where k is also arbitrarily chosen, there exist infinite numbers of points such A, with inversion A', such that

$$(OA).(OA') = k^2$$

and that all such infinite points lay on the line LL'. That is evident from the fact that the lengths of OA and OA' could be arbitrarily chosen such that the two lengths vary inversely, one increases, the other decreases, yet adhering to the equality: $(OA).(OA') = k^2$

Therefore, the inversion of a line LL' on any point, O, located on it is the same line.

Theorem II:

The inverse of a straight line with respect to a point **NOT** on the **same line** is a **circle**.

Proof

Given the line LL' with inversion origin at point O, not located on LL'.

59

Let points P be located on LL'.

Connect the points O and P, and define point P' as the inversion of point P.

Now, drop a perpendicular from the origin O on the line LL'. Name the point A the intersection of the perpendicular from O on LL'.

On the line OA, there is point A' that is the inversion of point A on O.

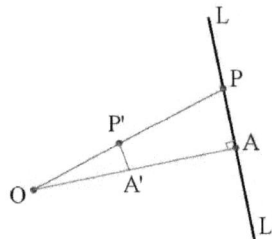

From the two properties of inversion, defined above, and from the above figure and the assumptions we made about the points A, A', P, P', and O. We conclude:

The products of the line lengths OA, OA', OP, and OP' satisfy the relationship:

OA. OA' = OP. OP' = k^2. ....(5.1)

i.e., the quadrilateral shape A'APP' must a **cyclic**, which means that all four vertices of the quadrilateral lay on a circle, such that the angle OP'A' = the angle OAP = 90°.

This can also be proven from the similar properties of the two triangles OP'A' and OAP, which share the same angle at O, each possesses a **right angle** (90°.) as follows:

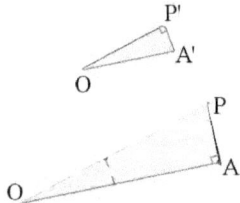

$$\frac{OP'}{OA} = \frac{OA'}{OP} = \frac{P'A'}{PA} \cdot \ldots\ldots\ldots\ldots\ldots\ldots\ldots(5.2)$$

From equation (5.2), equation (5.1) is obtained by using the properties of the cross multiplication of the terms of the ratios. OP'.OP = OA'.OA.

Therefore, the point P' traces a circle around O.

Note that k is the radius of inversion, not the radius of the locus of P' circle.

5.3. Inverse transformation of a circle on a point on same circle

Theorem III:

The inverse of a **circle** with respect to a point on its circumference is a **straight line**.

Proof

This is the reverse of the proof of Theorem II, above.

Given circle C to be inversed on point O on its periphery.

Draw a diameter from point O, passing by the origin of the circle, intersecting the circumference at point A.

A point P on the periphery of C makes the vertex of a right triangle with the points O and A.

Extend the line OA to an arbitrary point A', outside C, on the side of A, shown in the figure below.

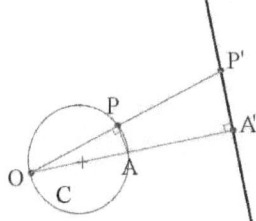

In order to prove that a line passing by the point A' will make the inversion of the circle C, we will extend the line OP to point P' such that the product OP'.OP should be equal to the product OA'.OA, of the already drawn line OAA'.

Connect the two extended points P' and A'. Using equation (5.2), above, we could prove that the triangle OA'P' is similar to the triangle OPA, such that the angle OPA = angle OA'P' = 90°.

Therefore, the line P'A' is the inverse of circle C on the point O, since all points on the periphery of C are inversed on the line P'A', with the condition OP'.OP = OA'.OA.

5.4. Inverse transformation of a circle on a point not on same circle

Theorem IV:

The inverse of a **circle** with respect to a point NOT on its circumference is a **circle**.

Proof

Assume that the origin O, of the x-y Cartesian coordinates, is the origin of inversion of the circle C_1:

$$x^2 + y^2 + 2gx + 2fy + c = 0. \quad \ldots\ldots\ldots\ldots\ldots\ldots\ldots(5.3)$$

Equation (5.3) can be written in the polar form by substituting by

$$y = r \sin \theta \quad \text{and} \quad x = r \cos \theta. \quad \ldots\ldots\ldots\ldots\ldots\ldots(5.4)$$

Thus, equation (5.3) becomes

$$r^2 + 2r(g \cos \theta + f \sin \theta) + c = 0. \quad \ldots\ldots\ldots\ldots\ldots\ldots(5.5)$$

Equation (5.5) makes it easy to change the polar distance r into $\dfrac{k^2}{r}$, which changes equation (5.5) into

$$\frac{k^4}{r^2} + 2 \frac{k^2}{r} (g \cos \theta + f \sin \theta) + c = 0. \quad \ldots\ldots\ldots\ldots\ldots\ldots(5.6)$$

Arranging the terms of equation (5.6), by multiplying it by r^2, we get

$$k^4 + 2k^2 r(g \cos \theta + f \sin \theta) + c r^2 = 0. \quad \ldots\ldots\ldots\ldots\ldots\ldots(5.7)$$

Again, using equation (5.4), in (5.7), we get the Cartesian form of the equation of the circle C_2:;

$$k^4 + 2k^2(gx + fy) + c(x^2 + y^2) = 0. \quad \ldots\ldots\ldots\ldots\ldots\ldots(5.8)$$

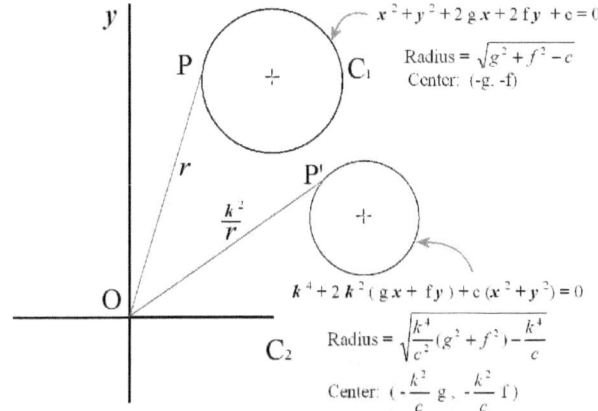

In order to prove that equation (5.8), of circle C_2, is the inversion of equation (5.3), of circle C_1, over the inversion point O, as both equations represent circles, we note that the product

$$(OP).(OP') = \frac{k^2}{r} \cdot r = k^2. \quad ....(5.9)$$

Therefore, the two circles C_1 and C_2, described by

Circle C_1, equation (5.3):

Radius $= \sqrt{g^2 + f^2 - c}$
Center: (-g, -f)

Circle C_2, equation (5.8):

Radius $= \sqrt{\dfrac{k^4}{c^2}(g^2 + f^2) - \dfrac{k^4}{c}}$

Center: $(-\dfrac{k^2}{c} g, -\dfrac{k^2}{c} f)$

represent inversion of each other on the outer point O.

5.5. Inverse transformation of angle between two curves

63

Theorem V:

The angle of intersection of two curves is unaltered by inversion.

Proof

Consider the two circles shown in the figure below, where one circle in the inverse of the other over the origin O.

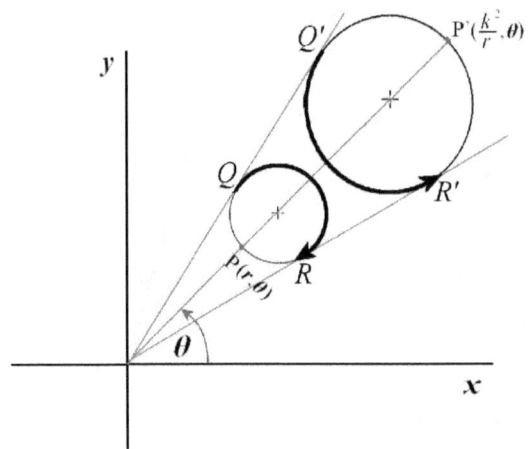

Point P' is the inverse of P, Q' inverse of Q, R' inverse of R. Therefore,

$(OP) . (OP') = (OQ) . (OQ')$

Thus, due to the similarities of the triangles shown in the figure,

angle OPQ = angle OQ'P'. ....(5.10)
(due to inversion, letters are reversed)

Also,

$(OP) . (OP') = (OR) . (OR')$
And
angle OPR = angle OR'P' . ...(5.11)
(also, due to inversion, letters are reversed)

Similarly,

angle QPR = angle Q'P'R' . ...(5.12)

Now, rotate the line OQQ' to coincide ultimately with the line OPP'. In the limit of rotation, the angle between the two tangents is equal to the angle between the two angles P'.

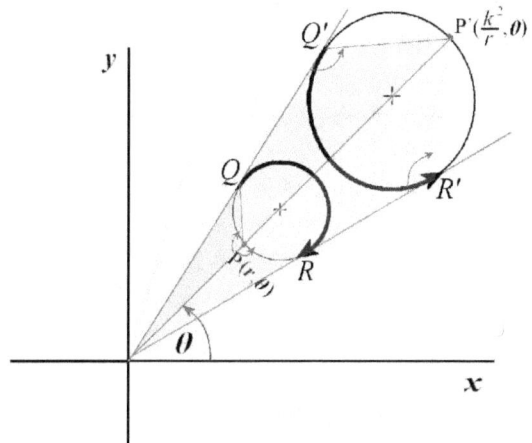

5.6. Operations involved in the inverse transformation of curves

Those operations are:

1. Translation
2. Magnification / Reduction
3. Rotation
4. Inversion
5. Reflection

1. Translation   2. Magnification/Reduction   3. Rotation

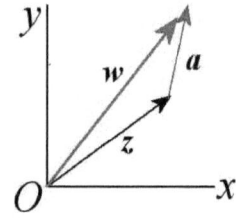 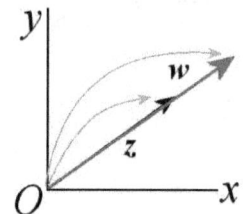 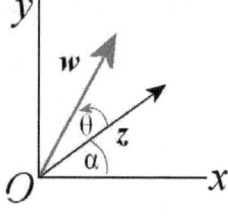

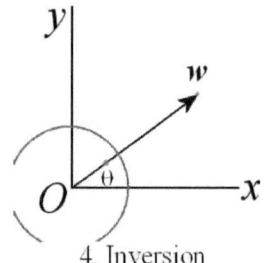 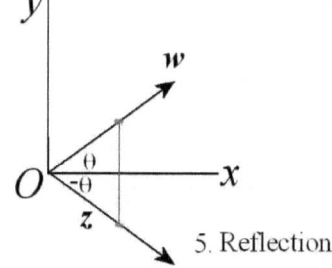

4. Inversion   5. Reflection

(1) Translation: $w = z + a$

Consider the source/sink mapping equation

$$w = \ln z = \ln r + i\theta. \quad\ldots\ldots\ldots\ldots\ldots\ldots\ldots(5.13)$$

Equation (3.13) is usually written with the strength constant $\dfrac{Q}{2\pi}$, where Q is the strength of the source or sink distributed over the circle with angle 2π. Therefore,

$$w = \frac{Q}{2\pi} \ln z \text{ Source of fluid (or radiation) . }\ldots\ldots\ldots\ldots\ldots\ldots(5.14)$$

$$w = -\frac{Q}{2\pi} \ln z \text{ Sink of fluid (or radiation) . }\ldots\ldots\ldots\ldots\ldots(5.15)$$

Equations (5.14) and (5.15) can be translated along the x axis by adding of subtracting the displacement constant, e.g., a,

$$w = \frac{Q}{2\pi} \ln (z + a) \quad \text{Source centered at } (a , 0)$$

$$w = \frac{Q}{2\pi} \ln (z - a) \text{Source centered at } (-a , 0)$$

The translation of complex variables is commonly used in fluid mechanics in formulating mapping of complex configurations, such as the flow of fluid past objects of various shapes and locations. For example, the flow of fluid past a wall and a round object is formulated by superposing the complex variables of the uniform flow, a circle, and a line. If a sink or source, such as fluid vortex as source or a draining outlet, then further complex variables of those objects are superposed on the initial formulation.

For example, the equation

$$w = \frac{Q_1}{2\pi} \ln (z + a) - \frac{Q_2}{2\pi} \ln (z - b) . \dots\dots\dots\dots\dots\dots\dots(5.16)$$

represents a source of strength Q_1, located at (a , 0) superposed on a sink of strength Q_2, located at (- b, 0). Equation (5.16) represents a **dipole or doublet**, commonly used in electrodynamics, thermodynamics, or fluid mechanics, where a source emits radiation and a sink drains it.

(2) Magnification / Reduction: $w = z\ a$

Where, a is scalar.

We have already shown that the multiplication of $\ln (z + a)$ by the scalar $\frac{Q_2}{2\pi}$ changes the strength (modulus or magnitude) of the complex variable. This is used to add the strength of flow of fluids, strength of a vortex, or the proportion of flow past an identified boundary.

(3) Rotation : $w = z\ e^{i\theta}$ (complex variable with unit modulus and argument)

Example 14

Changing the direction of flow

Given the complex variable for uniform flow,
$$\underline{w} = - U_o (x + i\, y)$$

Find the uniform flow directed at angle α with the above equation.

Solution

Let

$$\underline{w} = - U_0 (x + iy) \dots\dots\dots\dots\dots(5.17)$$

Where,

$$u = - U_0\ x \quad \text{and} \quad v = - U_0\ y$$

Since y is independent of x, therefore, equation (5.17) represents a **uniform flow** along the x-axis, with strength $|z| = U_0$.

Multiplying equation (5.17) by $e^{-i\alpha}$ results in a rotated uniform flow of the form.

$$\underline{w} = - U_0\ e^{-i\alpha}\ (x + iy)\dots\dots\dots\dots\dots(5.18)$$

Equation (5.18) can be written in the polar form as

$$\underline{w} = - U_0\ e^{-i\alpha}\ r\ (\cos \theta + i \sin\ \theta)\dots\dots\dots\dots\dots(5.19)$$
Thus,

$$\underline{w} = - U_0\ e^{-i\alpha}\ r\ e^{i\theta}\dots\dots\dots\dots\dots(5.20)$$

or,

$$\underline{w} = - U_0\ r\ e^{i(\theta - \alpha)}\dots\dots\dots\dots\dots(5.21)$$

Therefore, equation (5.21A4) represents **uniform flow directed** at angle α counter clockwise, with respect to x-axis.

Example 15

Circulating Vortex

Prove that the complex function

$$w = i\ ln\ z\dots\dots\dots\dots\dots(5.22)$$

represent circulating flow, or vortex.

Solution

We have shown in section 3.12 (Logarithmic Functions of complex numbers), that equation (5.22) can we written as follows.

$$w = u + iv = i\ ln\ z$$

and

68

$$z = e^{-iu+v} = e^v e^{-iu} = e^v (\cos u - i \sin u)$$

and
$$z = r(\cos u - i \sin u)$$
$$= x - i\,y$$

where,
$$r = e^v \text{ or } v = \ln r$$
$$r = \sqrt{x^2 + y^2} \text{ or } v = \ln \sqrt{x^2 + y^2} \ldots\ldots\ldots\ldots\ldots\ldots\ldots(5.23)$$
and
$$u = \theta + 2k\pi \text{ or } u = \tan^{-1}\frac{-y}{x} + 2k\pi. \ldots\ldots\ldots\ldots\ldots\ldots(5.24)$$

Equations (5.23) and (5.24) represent **circulating vortex** by virtue of pure angular dependence of u and the dependence of v on the radial length alone.

In order to graph equations (5.23) and (5.24), we will eliminate x or y separately as follow.

From (6), we get

$$x = \sqrt{e^v - y^2} \ldots\ldots\ldots\ldots\ldots\ldots(5.25)$$

Substituting by x from (5.25) into (5.24), we get a expression for u that could be graphed for various y = constant.

$$u = \tan^{-1}\frac{-y}{\sqrt{e^v - y^2}} + 2k\pi. \ldots\ldots\ldots\ldots\ldots(5.26)$$

Similarly, eliminating x between (5.23) and (5.24), we get a expression for v that could be graphed for various x = constant.

From (5.24), we get

$$y = x \tan(u - 2k\pi) \ldots\ldots\ldots\ldots\ldots\ldots(5.27)$$

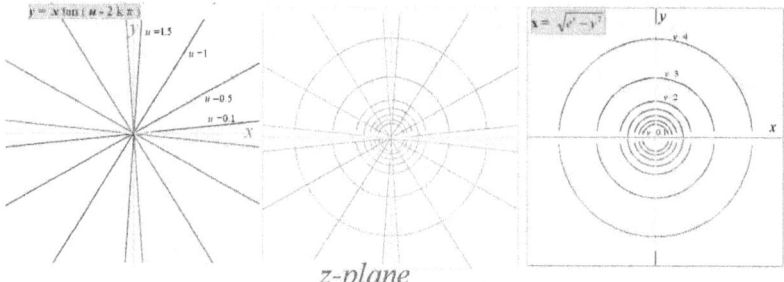

z-plane

Substituting by x from (5.27) into (5.23), we get a expression for v that could be graphed for various x = constant.

$$v = ln \sqrt{x^2 + x^2(\tan(u - 2k\pi))^2} \quad v = ln \ [x\sqrt{1 + (\tan(u - 2k\pi))^2}\ ]$$

$$v = ln \ [x.\sec(u - 2k\pi)] \dots\dots\dots\dots\dots\dots(5.28)$$

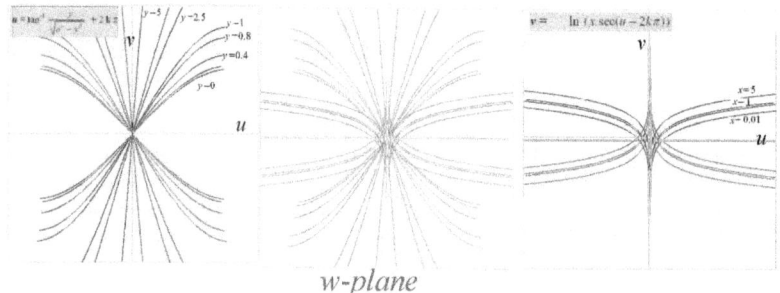

w-plane

Example 16

The Joukowski Problem

Superposing the following three complex functions constitute Joukowski fluid flow problem:

1. Uniform flow function:- $U_o\ z$

2. Circular cylinder function:- $U_o\ \dfrac{a^2}{z}$

3. Circulating Vortex: $i\dfrac{Q}{2\pi}\ ln\ \dfrac{z}{a}$

(5) Reflection: $w = \dfrac{1}{z}$

      We have already show that the mapping function $w = z$, represents **uniform linear flow** and that the **inversion of a line on a point not located on it is a circle**.

Example 17

Circular Cylinder Object

Prove that the complex variable of mapping function, $w_1 = \dfrac{1}{z}$, represents of a circular object with infinite length and inversion of $w_2 = z$.

Solution

$$u + i\,v = \frac{1}{x + iy} \quad ....(5.29)$$

$$= \frac{x - iy}{x^2 + y^2}$$

$$u = \frac{x}{x^2 + y^2} \quad ....(5.30)$$

$$v = -\frac{y}{x^2 + y^2} \quad ....(5.31)$$

$$x^2 + y^2 - x\,u = 0 ....(5.32)$$
$$x^2 + y^2 + y\,v = 0 ....(5.33)$$

Squaring and adding equations (5.30) and (5.31), we get

$$u^2 + v^2 = \frac{x^2}{(x^2 + y^2)^2} + \frac{y^2}{(x^2 + y^2)^2}$$

$$= \frac{x^2 + y^2}{(x^2 + y^2)^2} = \frac{1}{x^2 + y^2} = \frac{1}{r^2}$$

$$u^2 + v^2 = \frac{1}{r^2} ....(5.34)$$

Equations (5.32) and (5.33) represent circles for each value of u and v. Equation (5.34) is the inverse circle of the uniform flow $w = z$.

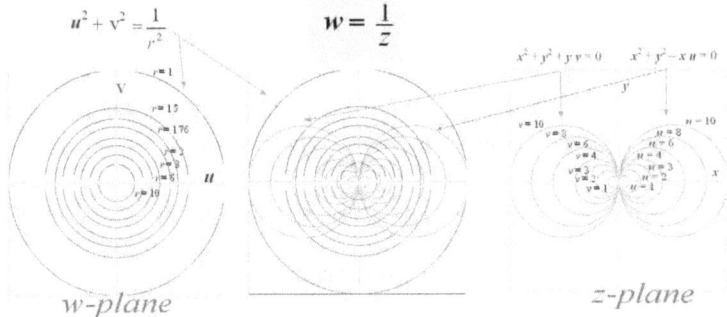

$$u^2 + v^2 = \frac{1}{r^2}$$

$$w = \frac{1}{z}$$

w-plane

z-plane

In the above figure, the circles in the z-plane are inversed in the w-plane. As the z-plane circles expand, the w-plane circles shrink, such that the condition of inversion, $\frac{1}{r}.r = k^2$, is satisfied.

The mapping function $w_1 + w_2 = z + \frac{1}{z}$, therefore, represents **uniform fluid flow past circular cylinder** and is one of the most famous mapping functions in fluid mechanics. By definition, a cylinder with infinite length is defined only by its radius, hence, the inversion circle was assumed such cylinder. In practice, cylinders have finite lengths, which requires more complicated computation in order to account for the irregular flow around two ends of the cylinder. Yet, in theoretical mathematical analysis, we pick the simple demonstration that exemplifies the basic principles of the subject, without the distracting details of practical applications.

Example 18

Uniform Flow Past Circular Cylinder Object

Graph and isolate the real and imaginary parts of the superposition the complex functions of **uniform flow**, z, and of **circular object**, $1/z$.

Solution

$$w = z + \frac{1}{z} \quad\dots\dots\dots\dots\dots\dots\dots(5.35)$$

Or,

$$u + iv = \frac{1}{x+iy} + x + iy$$

$$= \frac{x - iy}{x^2 + y^2} + x + iy \ldots\ldots\ldots\ldots\ldots\ldots(5.36)$$

Equating the real and imaginary parts of (5.36), we get

$$u = \frac{x}{x^2 + y^2} + x \ldots\ldots\ldots\ldots\ldots\ldots(5.37)$$

and

$$v = -\frac{y}{x^2 + y^2} + y \ldots\ldots\ldots\ldots\ldots\ldots(5.38)$$

Equations (5.37) and (5.38) give expressions for y in terms of x at u = constant and v = constant as follows:

$$x = \sqrt{-\frac{y}{v - y} - y^2} \ldots\ldots\ldots\ldots\ldots\ldots(5.39)$$

and

$$y = \sqrt{\frac{x}{u - x} - x^2} \ldots\ldots\ldots\ldots\ldots\ldots(5.40)$$

Graphing equations (5.39) and (5.40) show the famous **airfoil** profile which resulted from mapping the circular object placed in a uniform flow by conformal mapping using complex functions.

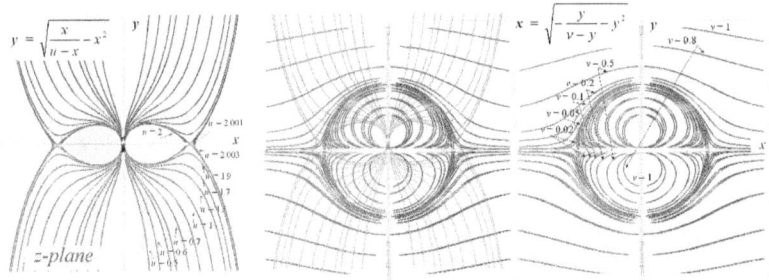

The contours of u = constant yield the **airfoil** profile. The contours of v = constant yield the grids of the network **orthogonal** on the airfoil profile. Clearly, the airfoil is attained by specific constants for u and v.

Cartesian Representation of $w\,(\,u\,,\,v\,)$:

73

Substituting by (5.39) in (5.37), we obtain an expression for u in terms of v at $y =$ constant as

$$u = \sqrt{2 + \frac{v - y}{y} - \frac{y}{v - y} - v^2} \quad \ldots\ldots\ldots\ldots\ldots\ldots\ldots\ldots\ldots(5.41)$$

Substituting by (5.40) in (5.38), we obtain an expression for v in terms of u at $x =$ constant as

$$v = \sqrt{-2 + \frac{u + x}{x} + \frac{x}{u - x} + 4x^2 \frac{u - x}{x} - u^2} \quad \ldots\ldots\ldots\ldots\ldots\ldots\ldots\ldots(5.42)$$

Graphing equations (5.41) and (5.42) gives the various values of x and y which satisfy the condition of creating circular object that can be mapped into airfoil.

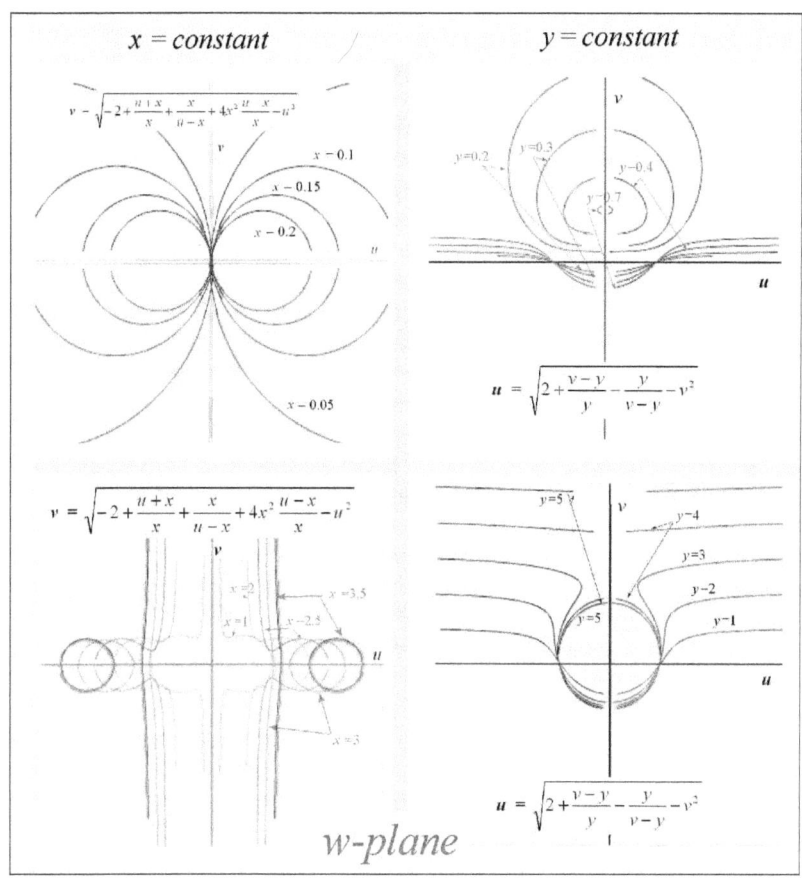

Graphing the complex function **w** for two different sets of **x**'s and **y**'s.

<u>Polar Representation of **w** (**u** , **v**):</u>

In the polar form, equations (5.37) and (5.36) could be written as

$$u = \frac{r\cos\theta}{r^2} + r\,\cos\theta \dots\dots\dots\dots\dots\dots(5.43)$$

$$v = -\frac{r\sin\theta}{r^2} + r\,\sin\theta \dots\dots\dots\dots\dots\dots(5.44)$$

Or

$$u = \cos\theta\left(\frac{1}{r}+r\right) \quad \text{.........................(5.45)}$$

$$v = \sin\theta\left(-\frac{1}{r}+r\right) \quad \text{.........................(5.46)}$$

Or

$$\cos\theta = \frac{u}{\frac{1}{r}+r} \quad \text{.........................(5.47)}$$

$$\sin\theta = \frac{v}{-\frac{1}{r}+r} \quad \text{.........................(5.48)}$$

Squaring equations (5.47) and (4.48) and adding, eliminates θ between the two equations and gives

$$1 = \left(\frac{u}{\frac{1}{r}+r}\right)^2 + \left(\frac{v}{-\frac{1}{r}+r}\right)^2 \quad \text{.........................(5.49)}$$

Equation (5.49) represent as set of ellipses varying on radii as r changes and are graphed below.

We could eliminate r between equations (5.45) and (5.46) as follows.

$$\frac{u}{\cos\theta} = \frac{1}{r}+r \text{.........................(5.50)}$$

$$\frac{v}{\sin\theta} = -\frac{1}{r}+r \text{.........................(5.51)}$$

Adding equations (5.50) and (5.51), we get

$$\frac{u}{\cos\theta} + \frac{v}{\sin\theta} = 2\,r \text{.........................(5.52)}$$

Subtracting equations (5.50) and (5.51), we get

$$\frac{u}{\cos\theta} - \frac{v}{\sin\theta} = 2\,\frac{1}{r}$$

Or

$$\left(\frac{u}{\cos\theta} - \frac{v}{\sin\theta}\right).\,r = 2 \text{.........................(5.53)}$$

Substituting by r from F(5.52) in (5.53) we get

$$(\frac{u}{\cos\theta} - \frac{v}{\sin\theta}) \cdot (\frac{u}{\cos\theta} + \frac{v}{\sin\theta}) = 4$$

Or

$$\left(\frac{u}{\cos\theta}\right)^2 - \left(\frac{v}{\sin\theta}\right)^2 = 4 ....(5.54)$$

Equation (5.54) represent as set of parabolas varying on radii as r changes. The dependence of w on θ and r is graphed as follows.

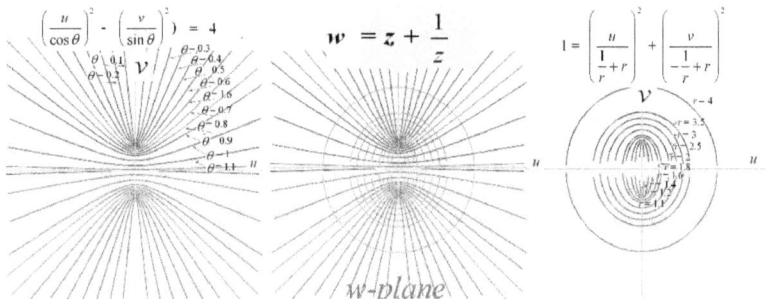

The uniform flow past circular cylinder is usually written as:

$$w = -U_o (z + \frac{a^2}{z}) ....(5.55)$$

Where, U_o is the strength of the fluid flow and a the radius of the circular cylinder.

Example 19

Circular Flow Past Circular Cylinder Object

Graph and isolate the real and imaginary parts of the superposition of the complex functions of **circular flow**, $i\ ln\ z$, and of **circular object**, $1/z$.

Solution

$$w = z + \frac{1}{z} + i\ ln\ z ....(5.56)$$

Substituting in equation (5.56) by $ln\ z = ln\ r + i\ \theta$, we get

Or,

77

$$u + iv = \frac{1}{x + iy} + x + iy + i(\ln r + i\theta)$$

$$= \frac{x - iy}{x^2 + y^2} + x + iy + i \ln r - \theta \dots\dots\dots\dots\dots\dots (5.57)$$

Equating the real and imaginary parts of (5.57), we get

$$u = \frac{x}{x^2 + y^2} + x - \theta \dots\dots\dots\dots\dots\dots (5.58)$$

and

$$v = -\frac{y}{x^2 + y^2} + y + \ln r \dots\dots\dots\dots\dots (5.59)$$

Cartesian Representation of $w\,(\,u\,,\,v\,)$:

The separation of x and y in equations (5.58) and (5.59) is impossible since the two equations are **transcendental** due to the presence of θ and $\ln r$ in the expressions of u and v. The only simplification we could get is as follows.

$$x = \sqrt{-\frac{y}{v - \ln r - y} - y^2} \dots\dots\dots\dots\dots (5.60)$$

and

$$y = \sqrt{\frac{x}{u + \theta - x} - x^2} \dots\dots\dots\dots\dots (5.61)$$

The graphing of equations (5.60) and (5.61) requires **numerical computation** where the two side of each transcendental equation are plotted separately, then the point of the intersection of the two sides, graphed separately, gives a single point on the z-plane. The process is repeated by varying v and u. In each graphing incident, a point is obtained on the z-plane.

Another approach in the solution of transcendental equations is through **numerical iteration** around initially guessed roots of the equation.

Graphic Solution of the Transcendental Equations (5.60) and (5.61):

Equation (5.60) can be written such that each side contains variables inseparable from the other side such as:

$$x^2 + y^2 = -\frac{y}{v - \ln r - y}$$

or

$$v - \frac{1}{2} \ln (x^2 + y^2) = -\frac{y}{x^2 + y^2} + y \ldots\ldots\ldots\ldots\ldots(5.62)$$

Each side of equation (5.62) could be equated to a constant C which will be determined graphically by plotting the curves of each side, finding their intersection coordinates, then determining the values of x and y that correspond to the given C, for each value of v. In the figure below, we gave v the values 1 and 0.5 in order to demonstrate how the intersection points could be determined, which correspond to the streamlines of $z(x, y)$. Thus,

$$-\frac{1}{2} \ln (x^2 + y^2) = C \ldots\ldots\ldots\ldots\ldots(5.63)$$

$$-\frac{y}{x^2 + y^2} + y = C \ldots\ldots\ldots\ldots\ldots(5.64)$$

From (5.63), we could write

$$y = \sqrt{e^{-2C} - x^2} \ldots\ldots\ldots\ldots\ldots(5.65)$$

$$x = \sqrt{\frac{y}{y - C} - y^2} \ldots\ldots\ldots\ldots\ldots(5.66)$$

We will give C arbitrary values, for each of those we will determine the x and y intersections of the two sides (5.63) and (5.64) which constitute a single point on the desired streamline contour in the z-plane.

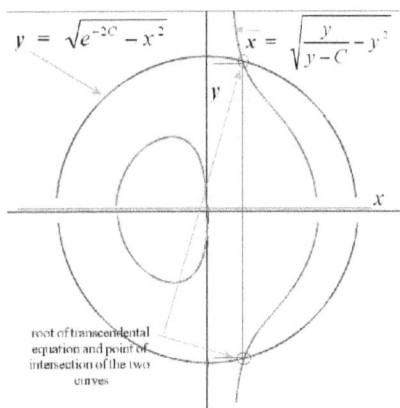

The remaining is a graphical determination as follows:

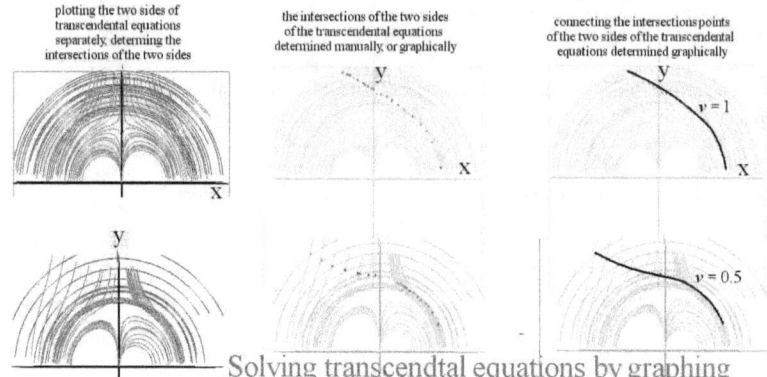

plotting the two sides of transcendental equations separately, determining the intersections of the two sides

the intersections of the two sides of the transcendental equations determined manually, or graphically

connecting the intersections points of the two sides of the transcendental equations determined graphically

Solving transcendtal equations by graphing

The above graphing is tedious and only used in demonstrating the essence of solution. Computational numerical iteration does that exactly through conditional loops that determine the points of interaction within determined or chosen level of precision. Matlab, FORTRAN, BASIC, or C programming are commonly used in graphing the airfoil of Joukowski with high speed computation and great ease.

connecting the intersections points
of the two sides of the transcendental
equations determined graphically

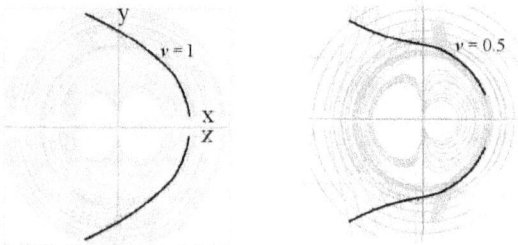

Polar Representation of $w\ (\ u\ ,\ v\ )$:

Writing equations (5.58) and (5.59) in the polar form, we get

$$u = \frac{\cos\theta}{r} + r\cos\theta - \theta \dots\dots\dots\dots\dots\dots(5.67)$$

and

$$v = -\frac{\sin\theta}{r} + r\sin\theta + \ln r \dots\dots\dots\dots\dots\dots(5.68)$$

80

Equations (5.67) and (5.68) are written as

$$\cos\theta = \frac{u+\theta}{\frac{1}{r}+r} \quad \ldots\ldots\ldots\ldots\ldots\ldots\ldots(5.69)$$

$$\sin\theta = \frac{v-\ln r}{-\frac{1}{r}+r} \quad \ldots\ldots\ldots\ldots\ldots\ldots\ldots(5.70)$$

Squaring equations (5.69) and (5.70) and adding, eliminates θ between the two equations and gives

$$1 = \left(\frac{u+\theta}{\frac{1}{r}+r}\right)^2 + \left(\frac{v-\ln r}{-\frac{1}{r}+r}\right)^2 \quad \ldots\ldots\ldots\ldots\ldots\ldots(5.71)$$

Equation (5.71) represent as set of ellipses varying on radii as r changes, translated along the u-axis as θ changes and along the v-axis as r changes.

We could eliminate r between equations (5.69) and (5.70) as follows.

$$\frac{u-\theta}{\cos\theta} = \frac{1}{r}+r \ldots\ldots\ldots\ldots\ldots\ldots(5.72)$$

$$\frac{v-\ln r}{\sin\theta} = -\frac{1}{r}+r \ldots\ldots\ldots\ldots\ldots\ldots(5.73)$$

Adding equations (5.72) and (5.73), we get

$$\frac{u-\theta}{\cos\theta} + \frac{v-\ln r}{\sin\theta} = 2r \ldots\ldots\ldots\ldots\ldots\ldots(5.74)$$

Subtracting equations (5.72) and (7.73), we get

$$\frac{u-\theta}{\cos\theta} - \frac{v-\ln r}{\sin\theta} = 2\frac{1}{r}$$

Or

$$(\frac{u-\theta}{\cos\theta} - \frac{v-\ln r}{\sin\theta}).\,r= 2 \ldots\ldots\ldots\ldots\ldots\ldots(5.75)$$

Substituting by r from (5.74) in (5.75) we get

$$\left(\frac{u-\theta}{\cos\theta} - \frac{v-\ln r}{\sin\theta}\right) \cdot \left(\frac{u-\theta}{\cos\theta} + \frac{v-\ln r}{\sin\theta}\right) = 4$$

Or

$$\left(\frac{u-\theta}{\cos\theta}\right)^2 - \left(\frac{v-\ln r}{\sin\theta}\right)^2 = 4 \dots\dots\dots\dots\dots(5.76)$$

Complex Velocity of Fluid Streaming

We will first rewrite equation (5.56) in the common form that show the strength of the fluid stream, U_o, the radius of the circular cylinder, a, and the strength of the vortex, $\frac{Q}{2\pi}$ (we already explained that the 2π rose from the symmetrical distribution of Q over the full angular range around the origin of the vortex $i \ln z$). Thus, equation (5.56) takes the form:

$$w = -U_o\left(z + \frac{a^2}{z}\right) + i\frac{Q}{2\pi}\ln z \dots\dots\dots\dots\dots(5.77)$$

Differentiating w with respect to z, equation (G16), gives the **complex velocity** as follows

complex velocity $= \dfrac{dw}{dz}$,

Thus

$$\frac{dw}{dz} - U_o\left(1 - \frac{a^2}{z^2}\right) + i\frac{Q}{2\pi}\cdot\frac{1}{z} \dots\dots\dots\dots\dots(5.78)$$

We have shown earlier that the magnification factor due to the conformal mapping transformation $w = f(z)$ is given by

$$\Delta w = \frac{dw}{dz}\Delta z$$

Hence, the **complex velocity** comprises the **magnification** or **reduction** of units of lengths in w-plane due to changes in the z-plane. Thus, the formulation in the z-plane of objects (circular cylinder), sinks, source, or vortex will result in changes in Δz, which in turn results changes Δw in the w-plane that can be gauged by the $\dfrac{dw}{dz}$.

The **maxima** and **minima** of equation (5.78) represent the stagnation and peak flow, respectively, as follows.

$$\frac{dw}{dz} - U_0\left(1 - \frac{a^2}{z^2}\right) + i\,\frac{Q}{2\pi}\cdot\frac{1}{z} = 0 \dots\dots\dots\dots\dots\dots(5.79)$$

Writing equation (5.79) in the quadratic form

$$Z^2(-U_0) + Z\left(i\,\frac{Q}{2\pi}\right) + (a^2 U_0) = 0 \dots\dots\dots\dots\dots\dots(5.80)$$

The roots, Z, of equation (5.80) are given by

$$Z = \frac{-i\dfrac{Q}{2\pi} \pm \sqrt{\left(i\dfrac{Q}{2\pi}\right)^2 - 4(-U_o)(a^2 U_o)}}{-2U_o} \dots\dots\dots\dots\dots\dots(5.81)$$

$$Z = i\frac{Q}{4\pi U_o} \pm \sqrt{a^2 - \left(\frac{Q}{4\pi U_o}\right)^2} \dots\dots\dots\dots\dots\dots(5.82)$$

The real and imaginary parts of the position of minimal/maximal complex velocity, Z, are determined with the value of the square root term

$$\sqrt{a^2 - \left(\frac{Q}{4\pi U_o}\right)^2} \dots\dots\dots\dots\dots\dots(5.83)$$

This could values of zero, real, or imaginary values as follows:

$$Z = i\frac{Q}{4\pi U_o} \pm \sqrt{a^2 - \left(\frac{Q}{4\pi U_o}\right)^2} \dots\dots\dots\dots\dots\dots(5.84)$$

Case 1: $a^2 = \left(\dfrac{Q}{4\pi U_o}\right)^2$

$$Z = X + i\,Y = i\frac{Q}{4\pi U_o}$$

$X = 0$

$$Y = \frac{Q}{4\pi U_o}$$

Case 2: $a > \left(\dfrac{Q}{4\pi U_o}\right)$

case 1

$$X = 0$$
$$Y = \frac{Q}{4\pi U_o}$$

case 2

$$X = \pm \sqrt{a^2 - \left(\frac{Q}{4\pi U_o}\right)^2}$$
$$Y = \frac{Q}{4\pi U_o}$$

case 3

$$X = 0$$
$$Y = \frac{Q}{4\pi U_o} \pm \sqrt{\left(\frac{Q}{4\pi U_o}\right)^2 - a^2}$$

$$Z = X + iY = i\frac{Q}{4\pi U_o} \pm \sqrt{a^2 - \left(\frac{Q}{4\pi U_o}\right)^2}$$

$$X = \pm \sqrt{a^2 - \left(\frac{Q}{4\pi U_o}\right)^2}$$

$$Y = \frac{Q}{4\pi U_o}$$

Case 3: $a < \left(\dfrac{Q}{4\pi U_o}\right)$

$$Z = X + iY = i\frac{Q}{4\pi U_o} \pm i\sqrt{\left(\frac{Q}{4\pi U_o}\right)^2 - a^2}$$

$$X = 0$$

$$Y = \frac{Q}{4\pi U_o} \pm \sqrt{\left(\frac{Q}{4\pi U_o}\right)^2 - a^2}$$

Hydrodynamic Force:

The **complex velocity** in equation (5.78) can be used to estimate the **fluid pressure** due to the interaction of the uniform or circulating flow with stationary objects, sinks, sources, or stationary walls.

In the case of point of mass of magnitude m, Newton's law of motion ($F = m\, \partial^2 x/\partial t^2$) gives the force applied on m as the product of m and the acceleration $\partial^2 x/\partial t^2$, where the latter measure the disturbance of the particles of m due to any external action.

The integration of the force; $F = m\, \partial^2 x/\partial t^2$, over the distance, where such acceleration took place, gives the **potential energy** (work done),

84

Work Done = F. $x = m\ \partial^2x/\partial t^2\ .x$..........................(5.85)

This applied also to $m\ g\ h$, for mass, m, gravitational acceleration, g, and height, h.

For example, Newton's law for a point particle of mass m, falling from height h, gives

½ m v² + m g h = constant..........................(5.86)

Which is the Newton's law of conservation of energy. The meaning of equation (5.86) is apparent in the case of ejecting a rocket upwards. At base, h = 0, but sufficient kinetic energy ½ m v² must be exerted in order to lift the rocket distance h in atmosphere.

Bernoulli expanded Newton's law of motion on multi-particle system by adding the **pressure** of particles to give

½ m v² + m g h + P V = constant ....(5.87)

The term PV accounts for the medium, either around the point mass m or that of m which affects the expenditure of the kinetic energy. In other words, ejecting a rocket in atmospheric pressure imposes more resistance than outer space vacuum or ejecting a rocket under water pressure, such as ejecting a rocket from a submerged submarine onto the outer atmosphere.

The reasoning behind adding the term PV in equation (5.87) can be deduced from the physics of gas dynamics. Here, gas pressure P is the force imparted per unit area due to the internal energy of the gas. Such gas pressure exerts work by expanding such unit area into unit volume such that

$$P\ .\ (\Delta A)\ .\ (\Delta d) = \frac{Force}{area}\ .\ \text{(unit area)}\ .\ \text{(unit distance)}$$

$$= \frac{Force}{area}\ .\ \text{(unit volume)}$$

$$= P.V$$

Therefore, the product PV, (or the product of pressure of gas times and the volume of gas), represents the **internal energy** that keeps the gas occupying its measurable volume with its measurable pressure. Therefore, all terms in equation (5.87) represent the **energy content** of the medium as any moment, given the constant value. This constant value is a rational formulation that implies that the energy content of a medium or system must be **conserved**. In other words, if the term PV increases, it must do so by taking energy either from the term ½ m v² or m g h or from both terms. The same applies to the other two terms. Thus energy cannot be created from void or wasted into void, but can be transformed between the three terms: ½ m v² , m g h , P V, with a constant net comprising the **Hamiltonian** of the system ,or its energy content.

However, the precise definition of the **Hamiltonian**, after Sir William Rowan Hamilton (1805-1865), is a mathematical function equal to the sum of the **kinetic** and **potential** energies of the system expressed in terms of the system's **coordinates** and **momenta** treated as independent variables.

Equation (5.87) is written after the substituting by unit volume, $V = 1$, and $\boldsymbol{m} = \rho$, density, such that

$$\tfrac{1}{2}\,\rho\,v^2 + \rho\,g\,\boldsymbol{h} + P = \text{constant} \dots\dots\dots\dots\dots\dots(5.88)$$

Equation (5.88) corresponds to the following entities

external kinetic energy + potential gravitational energy + internal energy = constant

　　　In theoretical application, the gravitational term, $\rho\,g\,\boldsymbol{h}$, is neglected unless we want to deal with specific application with gravitational forces acting on the system.

Hence, we get the famous Bernoulli's equation

$$\tfrac{1}{2}\,\rho\,v^2 + P = \text{constant} \dots\dots\dots\dots\dots\dots(5.89)$$

Where the constant C is determined by equating two systems

$$\tfrac{1}{2}\,\rho\,v^2 + P = \tfrac{1}{2}\,\rho_0\,v_0{}^2 + P_0 \dots\dots\dots\dots\dots\dots(5.90)$$

At $v_0 = 0$, equation (5.90) takes the form

$$P = P_0 - \tfrac{1}{2}\,\rho\,v^2 \dots\dots\dots\dots\dots\dots(5.91)$$

　　　Therefore, Bernoulli's law in equation (5.91) is a corollary of Newton's law that accounts for the mutual effects of particles. $P = 0$ signifies a point particle that cannot exert pressure on itself, as is the case in the basic Newton's law for point mass. In contrast, when dealing with compressible fluid, we use the ideal gas equation $PV = RT$, where R is the universal gas constant and T the absolute temperature of the system.

　　　If the fluid experiences **change of state**, such as freezing or evaporation of water, then the relationship between pressure P and density ρ gets complicated with the latent heats and transition of phases.

Equations (5.78) and (5.91) link the **complex velocity**, $\dfrac{dw}{dz}$, and the fluid pressure, P, through the Bernoulli equation, as follows.

$$P = P_0 - \tfrac{1}{2}\,\rho\left(\left|\frac{dw}{dz}\right|\right)^2 \dots\dots\dots\dots\dots\dots(5.92)$$

86

The force imparted on the circular cylinder, $\dfrac{a^2}{z}$, is obtained by integrating P, (5.92), over $\dfrac{1}{z}$ as follows.

In the x-direction, the element of force is: i P. dx
In the y-direction, the element of force is: P. dy

The term i P is the pressure directed in the imaginary y-axis.

$$\text{Fluid force} = \oint_{c=1/z}\left(P_o - \frac{1}{2}\left(\left|\frac{dw}{dz}\right|\right)^2\right).\,(\,dy + i\,dx\,)\ \dots\dots\dots\dots\dots\dots(5.93)$$

Rewriting equation (5.78) in polar form, we get

$$\frac{dw}{dz} = -U_o\left(1 - \frac{a^2}{r^2}e^{-2i\theta}\right) + i\,\frac{Q}{2\pi}\cdot\frac{1}{r}e^{-i\theta}\ \dots\dots\dots\dots\dots\dots(5.94)$$

$$= -U_o\left[1 - \frac{a^2}{r^2}(\cos 2\theta - i\sin 2\theta)\right] + i\,\frac{Q}{2\pi}\cdot\frac{1}{r}(\cos\theta - i\sin\theta)$$

$$= -U_o\left[1 - \frac{a^2}{r^2}\cos 2\theta\right] + \frac{Q}{2\pi r}\cdot\sin\theta + i\left[-U_o\,\frac{a^2}{r^2}\sin 2\theta + \frac{Q}{2\pi r}\cdot\cos\theta\right]$$

The modulus $\left|\dfrac{dw}{dz}\right| = \sqrt{(\text{Re})^2 + (\text{Im})^2}$

$$\left|\frac{dw}{dz}\right| = \sqrt{\left(-U_o(1-\frac{a^2}{r^2}\cos 2\theta) + \frac{Q}{2\pi r}son\theta\right)^2 + \left(-U_o\frac{a^2}{r^2}\sin 2\theta + \frac{Q}{2\pi r}\cos\theta\right)^2}$$

$$\left|\frac{dw}{dz}\right|^2 = U_o^2\left[1 - \frac{a^2}{r^2}\cos 2\theta + \left(\frac{a}{r}\right)^4\right] + \left(\frac{Q}{2\pi r}\right)^2 - \frac{QU_o}{\pi r}\sin\theta\,(1+\frac{a^2}{r^2})\dots\dots\dots\dots(5.95)$$

$$dy + i\,dx = d\,(r\sin\theta + i\,r\cos\theta)$$
$$= (\sin\theta\,dr + r\cos\theta\,d\theta) + i\,(-r\sin\theta\,d\theta + \cos\theta\,dr)\ \dots\dots\dots\dots\dots\dots(5.96)$$

Substituting equations (5.95) and (5.96) into (5.93), ignoring P_o, which is constant, we get

Real (Fluid force)

$$= \oint_{c=1/z} -\frac{1}{2}\cdot\left\{U_o^2\left[1 - \frac{a^2}{r^2}\cos 2\theta + \left(\frac{a}{r}\right)^4\right] + \left(\frac{Q}{2\pi r}\right)^2 - \frac{QU_o}{\pi r}\sin\theta\,(1+\frac{a^2}{r^2})\right\}(\sin\theta dr + r\cos\theta d\theta)$$

Imaginary (Fluid force)

$$= \oint_{c=1/z} -\frac{1}{2} \cdot \left\{ U_o^2 [1 - \frac{a^2}{r^2}\cos 2\theta + \left(\frac{a}{r}\right)^4] + \left(\frac{Q}{2\pi r}\right)^2 - \frac{QU_o}{\pi r} \sin\theta \, (1+\frac{a^2}{r^2}) \right\} (-r\sin\theta \, d\theta + \cos\theta \, dr)$$

5.7. Summary of bilinear transformation of lines and circles

The only operation that affects the nature of a curve is **inversion**. As we have shown, circles and straight lines are inverted to either straight lines or circles depending on the location of the origin of inversion.

In a bilinear transformation, combined transformations, such as translation, rotation, magnification, reduction, or inversion, are combined in one transformation such as

$$z = \frac{az+b}{cz+d}$$

Thus,

$$c\, z^2 + (d - a) - b = 0 ....(5.97)$$

Equation (5.97) is quadratic in z and is a bilinear transformation that maps any 3 points in the z-plane into another 3 points in the w-plane, such as:

$$\frac{(w - w_1)(w_2 - w_3)}{(w - w_3)(w_2 - w_1)} = \frac{(z - z_1)(z_2 - z_3)}{(z - z_3)(z_2 - z_1)}$$

CHAPTER 6

SERIES OF COMPLEX NUMBERS

A series of complex numbers takes the form

$$\sum_{i=1}^{n} a_i = a_1 + a_2 + a_3 + \ldots + a_n$$

Where the general term a_n is complex, .i.e.,

$$a_n = u_n + i\,v_n$$

6.1. Convergence of a series of complex numbers

The definition of convergence of a series of complex terms is the same as that of real terms. i.e., if take n terms of the form

$$A_n = a_1 + a_2 + a_3 + \ldots + a_n$$

If then if the A_n reaches a finite limit as n approach ∞ (infinity), then we say that the series $\sum_n a_n$ is convergent. Otherwise, it is divergent.

Let us write the series of complex terms as follows

$$\sum_n a_n = \sum_n u_n + i \sum_n v_n$$

The convergence of $\sum_n a_n$ implies the convergence of $\sum_n u_n$ and $\sum_n v_n$.

Absolute and conditional convergence:

1. A convergent series $\sum_n a_n$ is said to be absolutely convergent if $\sum_n |a_n|$ is convergent.

2. A convergent series $\sum_n a_n$ is said to be conditionally convergent if $\sum_n |a_n|$ is divergent.

We shall now show that is $\sum_n |a_n|$ is convergent, then $\sum_n a_n$ is also convergent.

Let $a_n = u_n + i\,v_n$

89

Then $|a_n| = \sqrt{u_n^2 + v_n^2}$

Now,

$0 \le u_n \le |a_n|$

$0 \le v_n \le |a_n|$

Since, $\sum_n |a_n|$ is convergent, then $\sum_n |u_n|$ and $\sum_n |v_n|$ are also convergent by the comparison test.

Now, from the series of real parts

$\sum_n u_n + i \sum_n v_n$ are convergent and hence

$\sum_n (u_n + iv_n)$ is convergent.

Therefore, $\sum_n a_n$ is convergent.

6.2. Power Series of complex variables

This is given by

$$\sum_{n=0}^{0} a_n z^n = a_0 + a_1 z + a_2 z^2 + \dots + a_n z^n + \dots$$

Where

$z = x + i\,y$

and

$a_n = u_n + i\,v_n$

We shall now show that if $\sum_{n=0}^{0} a_n z^n$ is convergent for $z = z_0$, then $\sum_{n=0}^{0} a_n z^n$ is absolutely convergent for all z satisfying the condition $|z| < |z_0|$. The geometrical meaning of this is that is $\sum_{n=0}^{0} a_n z^n$ is convergent for some points $z = z_0$ then it is absolutely convergent for all points inside the circle passing through the point $z = z_0$.

90

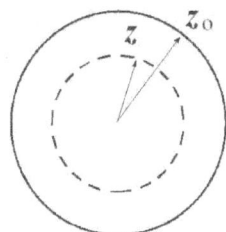

Since $\sum_{n=0}^{0} a_n z^n$ is convergent at $z = z_0$, thus $\sum_{n=0}^{0} a_n z_o^n$ is convergent. Then $a_n z_o^n \rightarrow 0$ as n $\rightarrow \infty$, hence

$|a_n z_o^n| < 1$ for all n > n$_o$

$$\left| \sum_{n=0}^{0} a_n z^n \right| = \left| a_n z_o^n \left(\frac{z}{z_o} \right)^n \right| < \left| \frac{z}{z_o} \right|^n$$

For all n > n$_o$ and for all points inside the circle $|z| < z_0$, i.e., $\left| \frac{z}{z_o} \right| < 1$.

Hence, $\sum_{n} \left| \frac{z}{z_o} \right|^n$ is convergent, being a geometric series with common ratio less than unity.

Therefore,

$$\sum_{n=0}^{0} \left| a_n z^n \right| \text{ is convergent}$$

and

$$\sum_{n=0}^{0} a_n z^n \text{ is absolutely convergent}$$

6.3. Circle of convergence

This has the following properties:

1. The series $\sum_{n=0}^{0} a_n z^n$ is absolutely convergent for all points inside the circle.

2. The series is not convergent for any point outside the circle.

3. The series may or may not be convergent on the periphery of the circle itself.

The radius of this circle is called the **radius of convergence**. This determined in a manner similar to that of the **ratio test**.

91

The following rule can be adapted:

Find the limit of the ratio $\dfrac{a_{n+1}}{a_n}$; or $\lim\limits_{n\to\infty}\left|\dfrac{a_{n+1}}{a_n}\right|$.

Denote that limit by l; or $\lim\limits_{n\to\infty}\left|\dfrac{a_{n+1}}{a_n}\right| = l$.

The radius of convergence is then $\dfrac{1}{l}$.

Example 20

Consider the exponential series

$$e^z = 1 + z + \frac{z^2}{2!} + \ldots + \frac{z^n}{n!} +$$

Solution

$$a_n = \frac{1}{n!}$$

$$a_{n+1} = \frac{1}{(n+1)!}$$

$$\lim_{n\to\infty}\left|\frac{a_{n+1}}{a_n}\right| = \lim_{n\to\infty}\frac{n!}{(n+1)!}$$

$$= \lim_{n\to\infty}\frac{n!}{(n+1).n!}$$

$$= \lim_{n\to\infty}\frac{1}{n+1} = 0$$

As $n \to \infty$, $l \to 0$, radius of convergence $= \infty$. Thus, the exponential series is absolutely convergent for all points in the z-plane.

Example 21

Consider the geometric series

$$1 + z + z^2 + z^3 + \ldots + z^n +$$

Solution

92

Consider the geometric series $1 + z + z^2 + z^3 + ... + z^n + ...$

$$\lim_{n \to \infty} \left| \frac{a_{n+1}}{a_n} \right| = 1 \text{, hence the radius of convergence} = 1.$$

Thus,

z^n does not tend to zero as n $\to \infty$.

Thus, the geometric series is absolutely convergent for all points inside the unit circle. It is not convergent for any point outside the unit circle. For points on the circle itself, $| z^n | = 1$.

Example 22

$$\text{Consider the series } \sum_{n=0}^{0} \frac{z^n}{n^2}$$

Solution

$$\lim_{n \to \infty} \left| \frac{a_{n+1}}{a_n} \right| = \lim_{n \to \infty} \frac{n^2}{(n+1)^2}$$

$$= \lim_{n \to \infty} \frac{1}{(1 + \frac{1}{n})^2} = 1 \text{, hence the radius of convergence} = 1.$$

For all points on the circle $\left| \frac{z^n}{n^2} \right| = \frac{1}{n^2}$ and we know that $\sum_n \frac{1}{n^2}$ is convergent.

6.4. Line Integral of a function of complex variables

Let C be a curve represented parametrically by the equations

$x = \varphi(t)$
$y = \psi(t)$

where, φ and ψ are **differentiable** functions of the real variable t. Let $f(z)$ be a **single-valued continuous** function of z for all points on the curve C.

Subdivide the arc C into large number of small subdivisions $z_0, z_1,..., z_n$. In each subdivision, take any point, ζ_i on the arc on the subdivision z_i and z_{i+1}.

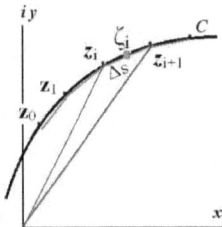

From the coordinates ζ_i, form the sum

$$\sum_{i=1}^{n} f(\zeta_i).(z_i - z_{i+1}.)$$

Upon taking the limit of the above some as n $\rightarrow \infty$ in such a way that the arcs are indefinitely small, then the limit of the sum becomes the integral:

$$\lim_{n \leftarrow \infty} \sum_{i=1}^{n} f(\zeta_i).(z_i - z_{i+1}.) = \oint_c f(z).dz$$

Then suppose that

$$f(z) = u(x,y) + iv(x,y)$$
$$z = x + iy$$
$$dz = dx + idy$$

Then

$$\oint_c f(z).dz = \oint_c (u + iv).(dx + idy)$$
$$= \oint_c (u\,dx - v\,dy) + i\oint_c (v\,dx + u\,dy)$$

Thus, we see that the line integral of a function of complex variables depends on the curvilinear integrals of real functions.

The conjugate functions u and v were proven earlier to satisfy **Laplace's equation** by virtue of the differentiability, continuity, and single-value of the function $f(z) = u(x,y) + iv(x,y)$. We will also show that conjugate functions u and v satisfy **Cauchy-Riemann equation**.

CHAPTER 7

CAUCHY'S THEOREM

Let C be a simple closed contour or a curve that does not cross itself.
Let $f(z)$ be a regular or analytic function in a domain containing C together with its interior.

Prove that $\oint_c f(z).dz = 0$

Proof

$$\oint_c f(z)dz = \oint_c (u + iv)(dx + idy)$$

$$= \oint_c (u\,dx - v\,dy) + i \oint_c (v\,dx + u\,dy) \dots\dots\dots\dots\dots(7.1)$$

We will first use Green's Theory to convert the line-integral in equation (7.1) into surface-integral as follows.

In the diagram, below, of curve C, incremental distances Δx and Δy, and surface area element Δs, the unit normal $\bar{n}$ that is perpendicular on Δs is related to the unit vector dr along Δs by the relations:

$$d r = dx + idy \dots\dots\dots\dots\dots(7.2)$$
where
$$ds = \sqrt{(dx)^2 + (dy)^2} \dots\dots\dots\dots\dots(7.3)$$
and
$$\bar{n}\,ds = dy - idx \dots\dots\dots\dots\dots(7.4)$$

Equation (7.4) is obtained from (7.2) by the trigonometric property

(slope of dr) . (slope of $\bar{n}$. ds) = -1

Hence,

$$(\text{slope of } \bar{n}\,ds) = -1/(\text{slope of } dr) = -\frac{1}{slopeOf(dr)} = \frac{-1}{\dfrac{dy}{dx}}$$

Consider the function

$$f(u, y) = u + iv \dots\dots\dots\dots\dots(7.5)$$

From (7.4) and (7.5), we get

$$f(u,y). \bar{n}\,ds = (u + iv).(dy - idx)$$

95

Thus

$$\oint_c f(z)dz = \oint_c (u\,dx - v\,dy) + i\oint_c (v\,dx + u\,dy) \dots\dots\dots\dots\dots\dots(7.6)$$

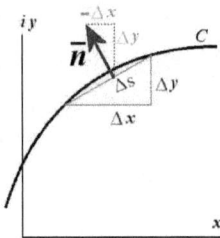

$$\oint_c f(z)dz = \iint_S f(u,y).\,\bar{n}\,ds \dots\dots\dots\dots\dots\dots(7.7)$$

The surface-integral in equation (7.7), which corresponds to the line-integral, is replaced by Green's Theorem as follows:

$$\oint_c (u\,dx - v\,dy) = \iint_S \left(-\frac{\partial v}{\partial x} - \frac{\partial u}{\partial y}\right) du\,dy \dots\dots\dots\dots\dots\dots(7.8)$$

$$\oint_c (v\,dx + u\,dy) = \iint_S \left(\frac{\partial u}{\partial x} - \frac{\partial v}{\partial y}\right) du\,dy \dots\dots\dots\dots\dots\dots(7.9)$$

From Cauchy-Riemann and Laplace equations we get

$$\frac{\partial u}{\partial x} = \frac{\partial v}{\partial y} \dots\dots\dots\dots\dots\dots(7.10)$$

$$\frac{\partial u}{\partial y} = -\frac{\partial v}{\partial x} \dots\dots\dots\dots\dots\dots(7.11)$$

Equations (7.10) and (7.11) implies that the integrals of equations (7.8) and (7.9) vanish

Thus,

$$\oint_c f(z)dz = 0$$

Corollary of Cauchy's Theorem:

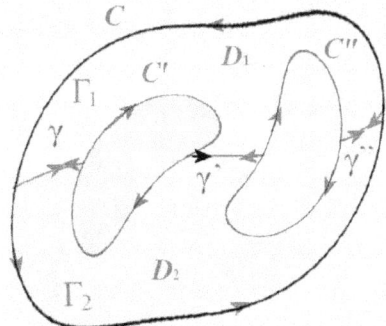

Let C be a closed contour, enclosing a finite number of the closed contours C', C", ..., as shown the above figure.

Let $f(z)$ be a regular function in a domain containing C, C', C" and the area between them.

Thus,

$$\oint_C f(z)dz = \oint_{C'} f(z)dz + \oint_{C''} f(z)dz + ... = z$$

Divide the area inside C into the two subdomains D_1 and D_2, by a set of curves γ, γ', and γ". Let the bounding contours of the subdomains D_1 and D_2, be Γ_1 and Γ_2,.

Since $f(z)$ is regular in each of the subdomains D_1 and D_2, then

$$\oint_{\Gamma_1} f(z)dz = 0$$

$$\oint_{\Gamma_2} f(z)dz = 0$$

Hence, according to Cauchy's theory,

$$\oint_{\Gamma_1} f(z)dz + \oint_{\Gamma_2} f(z)dz = 0$$

The integrals on the segments γ, γ', and γ" cancel one another, as the enter twice in opposite directions, i.e., with opposite signs.

Hence,

$$\oint_C f(z)dz = \oint_{C'} f(z)dz + \oint_{C''} f(z)dz + ... = 0$$

97

7.1. Isolating singularities with Cauchy's problems

Example 23

Evaluate $\displaystyle\oint_C \frac{dz}{z}$,

where C is a closed contour enclosing the origin of the function in its interior.

Solution

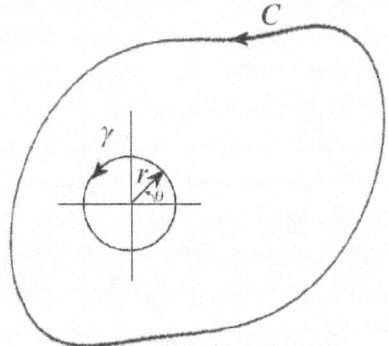

Since the function $\dfrac{1}{z}$ has a singularity at $z = 0$, we cannot apply Cauchy's theory without isolating the origin by drawing a circle γ around the center $z = 0$, with radius r, and lies within C.

Hence, by applying Cauchy's theorem on the new configuration, we get

$$\oint_C \frac{dz}{z} = \oint_\gamma \frac{dz}{z} \quad \ldots\ldots\ldots\ldots\ldots\ldots(7.12)$$

On the circle γ, we have

$$z = r e^{i\theta} \quad \ldots\ldots\ldots\ldots\ldots\ldots(7.13)$$

$$dz = i r e^{i\theta} d\theta \ldots\ldots\ldots\ldots\ldots\ldots(7.14)$$

Therefore,

$$\oint_C \frac{dz}{z} = \int_0^{2\pi} \frac{1}{re^{i\theta}}\, i r e^{i\theta} d\theta$$

98

$$= \int_0^{2\pi} i \, d\theta$$
$$= 2\pi i$$

Example 24

Evaluate $\oint_C \dfrac{(z^2+1)dz}{z}$,

where C is a closed contour enclosing the origin of the function in its interior.

Solution

We have $\oint_C \dfrac{(z^2+1)dz}{z} = \oint_C z\,dz + \oint_C \dfrac{dz}{z}$

The first integral, $\oint_C z\,dz$, vanishes by Cauchy's Theorem, if z in being regular function.

The second integral, $\oint_C \dfrac{dz}{z}$, can be evaluated as the previous example by drawing a circle γ, with center at the origin of the function. Thus,

$$\oint_C \dfrac{(z^2+1)dz}{z} = 2\pi i$$

7.2. Poles of singular functions

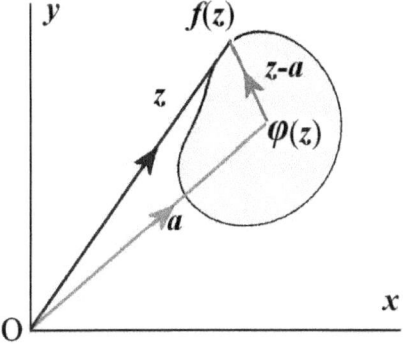

Poles:

Suppose that we have a function $f(z)$ regular in a certain circular domain except at the center (a) of the circle, shown in the above figure.

Suppose that $f(z)$ can be expanded in the neighborhood of $z = a$ in the form

$$f(z) = \frac{C_n}{(z-a)^n} + \frac{C_{n-1}}{(z-a)^{n-1}} + \ldots + \frac{C_n}{(z-a)^n} + \Phi(z)$$

7.2.1. Poles of simple order

In the above expansion of $f(z)$, $\Phi(z)$ is regular at $z = a$.

Here, we have a pole at $z = a$ of order (n),
$n \geq 1$ and
C_1 is the residue,
$C_n \neq 0$, and

$$\oint_C \Phi(z) = 0$$

Case I:

Suppose that we have a simple pole, of the first order, at $z = a$.
Then, if
$$f(z) = \frac{C_1}{z-a} + \Phi(z)$$
Therefore,

$$(z-a)f(z) = C_1 + (z-a)\,\Phi(z)$$

Let $z \rightarrow a$

Therefore, the 1st order pole corresponds to:

$$C_1 = \lim_{z \rightarrow a} (z-a)\,[f(z) - \Phi(z)] = \lim_{z \rightarrow a} (z-a)f(z)$$

7.2.2. Poles of nth order

Case II:

Suppose that we have a pole of order (n) at $z = a$.
Write the $f(z)$ in the form

$$f(z) = \frac{\Phi(z)}{(z-a)^n}$$

Where $\Phi(z)$ consists of the numerator and all the factors of the denominator <u>except</u> the factor $(z - a)^n$.

Now, let us expand $\Phi(z)$ by Taylor's theory in the neighborhood of $z = a$.

Therefore,

$$\Phi(z) = \Phi(a + \overline{z - a})$$

$$= \Phi(a) + \Phi'(a)(z-a) + \frac{(x-a)^2}{2!}\Phi''(a) + ..$$

$$+ \frac{(x-a)^{n-1}}{(n-1)!}\Phi^{n-1}(a) + \frac{(x-a)^n}{n!}\Phi^n(a)$$

Dividing by $(z - a.)^n$ we get.

$$\frac{\Phi(z)}{(z-a)^n} = \frac{\Phi(a)}{(z-a)^n} + \frac{\Phi'(a)}{(z-a)^{n-1}} + \frac{\Phi''(a)}{2!(z-a)^{n-2}} + ..+ \frac{\Phi^{n-1}(a)}{(n-1)!(z-a)} + \frac{\Phi^n(a)}{n!}$$

$$f(z) = \frac{\Phi(a)}{(z-a)^n} + \frac{\Phi'(a)}{(z-a)^{n-1}} + \frac{\Phi''(a)}{2!(z-a)^{n-2}} + ..+ \frac{\Phi^{n-1}(a)}{(n-1)!(z-a)} + \frac{\Phi^n(a)}{n!}$$

Since, first term in the expansion of $f(z)$; $\dfrac{\Phi(a)}{(z-a)^n}$, starts at zero differential exponent of $\Phi(a)$, therefore, the <i>n</i><u>th</u> order pole corresponds to:

$$C_n = \frac{\Phi^{n-1}(a)}{(n-1)!}$$

Since for convenience we used the differential exponent on $\Phi(a)$, we will clarify our notation as follows:

$$C_n = \frac{1}{(n-1)!}\left[\frac{d^{n-1}}{dz^{n-1}}\Phi(z)\right]_{z=a}$$

7.2.3. Residues at the <i>n</i><u>th</u> pole

Where, $f(z)$ is some regular function, the residue at $z = a$ is

$$\text{Residue} = \frac{1}{(n-1)!}\left[\frac{d^{n-1}}{dz^{n-1}}\Phi(z)\right]_{z=a}$$

Example 25

Find the poles of $\frac{z+1}{z}$

Solution

$$\frac{z+1}{z} = 1 + \frac{1}{z}$$

Here we have a simple pole of first order and residue 1, which is the numerator of the term $\frac{1}{z}$.

Example 26

Find the poles of $\frac{e^z}{z}$

Solution

$$\frac{e^z}{z} = \frac{1}{z}\left(1 + z + \frac{z^2}{2!} + \dots\right)$$

$$= \frac{1}{z} + 1 + \frac{z}{2!} + \dots$$

$$= \frac{1}{z} + \Phi(z)$$

Where $\Phi(z)$ is regular.

Here we have origin at $z = 0$, a simple pole of residue 1, which is the numerator of the term $\frac{1}{z}$.

Example 27

Find the poles of

$$\frac{1}{z(z-1)^2}$$

Solution

Here, we have two poles:
one of order 1, at $z = 0$ and
a pole of order 2, at $z = 1$.

Therefore,

Pole $z = 0$ → Residue = $\displaystyle\lim_{z \to 0} \frac{1}{z(z-1)^2} = 1$

Pole $z = 1$ → Residue = $\displaystyle\frac{1}{(2-1)!}\left[\frac{d}{dz}\frac{1}{z}\right]_{z=1} = \left[-\frac{1}{z^2}\right]_{z=1} = -1$

Where $\Phi(z)$ is regular.

Here we have origin at $z = 0$, a simple pole of residue 1, which is the numerator of the term $\dfrac{1}{z}$.

<u>Another Method of Solution</u>

The residue at the pole $\mathbf{z} = 1$, which is 2nd order, can also be evaluated by replacing z-1 by ζ, using the binomial series to expand $\dfrac{1}{1+\xi}$, as follows.

$$\frac{1}{z(z-1)^2} = \frac{1}{\xi^2(1+\xi)}$$

$$= \frac{1}{\xi^2}(1 - \zeta + \zeta^2 - \zeta^3 \dots)$$

$$= \frac{1}{\xi^2} - \frac{1}{\xi} + \Phi(\zeta)$$

$$= \frac{1}{(z-1)^2} - \frac{1}{z-1} + \Psi(z)$$

Where $\Psi(z)$ is regular at $z = 1$ and the residue is -1, the numerator of the term $\dfrac{1}{z-1}$.

7.3. Cauchy's Theorem of Residue

The **line-integral** of a function, that is regular in a domain D, which is enclosed by a closed contour C and contains a finite numbers of poles $a, a', a'', ..,$ in its interior, is given by:

$$\oint_C f(z)\, dz = 2\,\pi\, i \,.\, ((\text{sum of residues of} f(z) \text{ at its poles inside C}))$$

Consider the poles of $f(z)$ at $z = 0$ and suppose that $f(z)$ can be expanded around a in the form:

$$f(z) = \frac{C_n}{(z-a)^n} + \frac{C_{n-1}}{(z-a)^{n-1}} + \dots\dots + \frac{C_1}{z-a} + \Phi(z)$$

Where $\Phi(z)$ is regular in the same circular domain, with center a and radius r, and similarly for the poles $a, a', a'', ..,$ etc.

Now, by taking $r, r', r'', ..,$ small enough, we can ensure that they do not overlap and that they all lie inside C.

Draw circle γ, of radius ρ, around a as a center (a is complex variable, not a scalar, see figure) such that $\rho < r$. Further, draw similar circles around the remaining poles.

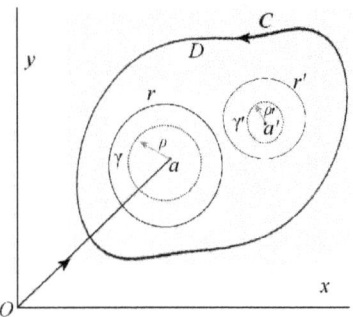

According to Cauchy's Theorem Corollary

$$\oint_C f(z)dz = \oint_\gamma f(z)dz + \oint_{\gamma'} f(z)dz + \dots$$

Where all integrals are with positive signs, i.e., counter clockwise.

Now, consider $\oint_\gamma f(z)dz$

$$\oint_\gamma f(z)\,dz = \sum_{m=1}^{n} C_m \int_\gamma \frac{dz}{(z-a)^m} + \int_\gamma \Phi(z)dz$$

Now, the last integral, $\int_\gamma \Phi(z)dz = 0$ since $\Phi(z)$ is regular for all points on the circle γ.

As illustrated in the figure, below, we can put

$$z = \rho\, e^{i\theta} + a$$
$$dz = i\rho\, e^{i\theta}\,d\theta$$

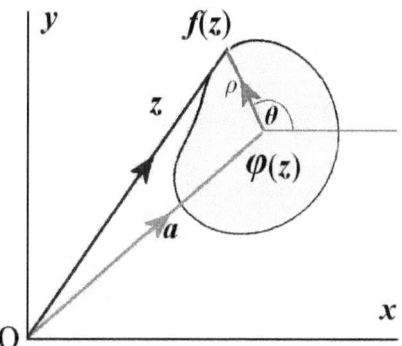

$$\oint_\gamma \frac{dz}{(z-a)^m} = \int_{-\pi}^{\pi} \frac{i\rho e^{i\theta}\,d\theta}{\rho^m e^{im\theta}}$$

$$= \frac{i}{\rho^{m-1}} \frac{\left[e^{i(1-m)\theta}\right]_{-\pi}^{\pi}}{i(1-m)}$$

$$= \frac{1}{\rho^{m-1}(1-m)}\left[e^{i(1-m)\pi} - e^{-i(1-m)\pi}\right] = 0$$

Since, $e^{i\theta} - e^{-i\theta} = 2\,i \sin\theta$

When m = 1 (one pole of first order), then

$$\oint_\gamma f(z)\,dz = C_1 \oint_\gamma \frac{dz}{z-a} = C_1.\,2\pi i$$

Because,

$$\oint_{\gamma} f(z)\, dz = \int_{-\pi}^{\pi} \frac{ipe^{i\theta} d\theta}{pe^{i\theta}} = i\int_{-\pi}^{\pi} d\theta = 2\pi i$$

Similarly, for integrals around γ', γ'', ..

$$\oint_{C} f(z)\, dz = 2\pi i\, (C_1 + C_2 + C_3 + \ldots)$$

$$= 2\pi i\, (\text{sum of residues})$$

7.3.1. Examples on Cauchy's Theorem of Residues

Example 28

Evaluate

$$\int_{C} \frac{5z-2}{z(z-1)}\, dz$$

Where $|z| = 2$ and the integral describes counter clock-wise path.

Solution

Here, we have two simple poles, which lie inside C which has radius = 2. The two poles are

$z = 0$

$z = 1$

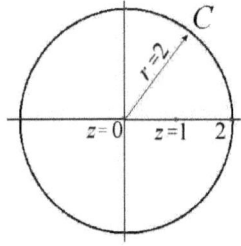

The residues of the two poles are

$$z = 0 \rightarrow \text{Residue} = \lim_{z \to 0} \frac{(5z-2)}{z(z-1)} = 2$$

$$z = 1 \rightarrow \text{Residue} = \lim_{z \to 1} (z-1)\frac{(5z-2)}{z(z-1)} = 3$$

106

Therefore,

$$\int_C \frac{5z-2}{z(z-1)} dz = 2\pi i\ (\ 2 + 3\ )$$
$$= 10\ \pi\ i$$

We can also resolve the given fraction into its partial terms as follows:

$$\int_C \frac{5z-2}{z(z-1)} dz = \int_C \frac{2}{z} dz + \int_C \frac{3}{(z-1)} dz$$
$$= 2(2)\ \pi\ i + 2(3)\ \pi\ i$$
$$= 10\ \pi\ i$$

Example 29

Find the value of the integral

$$\int_C \frac{dz}{z^3(z+4)}$$

Where the circle C is defined by
(i) $|z| = 2$
(ii) $|z+2| = 3$
(iii) $|z-4| = 1$

Solution

Here, we have

simple pole at $z = 0$, and

3rd order pole at $z = -4$

Therefore,

$$z = 0 \rightarrow \text{Residue} = \frac{1}{2!}\left[\frac{d^2}{dz^2}\left(\frac{1}{z+4}\right)\right]_{z=0}$$
$$= \frac{1}{2!}\left[\frac{(-1)(-2)}{(z+4)^3}\right]_{z=0} = \frac{1}{64}$$

$z = -4 \rightarrow$ Residue $= \lim_{z \to -4} (z+4) \dfrac{1}{z^3(z+4)} = -\dfrac{1}{64}$

Another method of solution

$$\frac{1}{z^3(z+4)} = \frac{1}{4z^3(1+\frac{z}{4})}$$

$$= \frac{1}{4z^3}[\,1 - \frac{z}{4} + \frac{z^2}{16} \ \ldots\ldots]$$

$$= \frac{1}{4z^3} - \frac{1}{16z^2} + \frac{1}{64z} \ \ldots\ldots$$

Therefore, the numerator of the term $\dfrac{1}{z}$, which is $\dfrac{1}{64}$, is the residue the pole $z = 0$.

Evaluation of the integral at different poles:

(i) $|z| = 2$

Here, the two poles lie on the circle C.

$$\rightarrow \int_C \frac{dz}{z^3(z+4)} = 2\pi i \frac{1}{64}$$

$$= \frac{i\pi}{32}$$

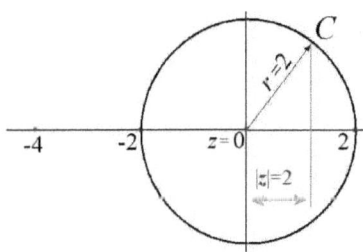

(i) $|z| = 2$

(ii) $|z+2| = 3$ Here, the two poles lie inside the circle C.

$\rightarrow \int\limits_{c} \dfrac{dz}{z^3(z+4)} = 2\pi i\left(-\dfrac{1}{64}+\dfrac{1}{64}\right) = 0$

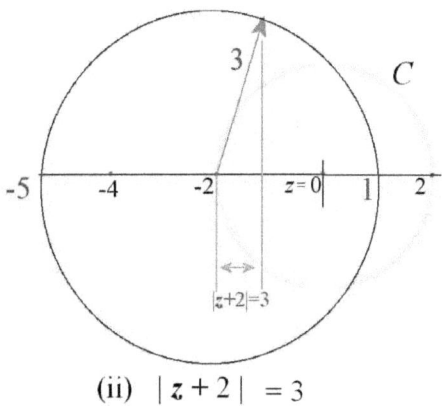

(ii) $|z+2| = 3$

(iii) $|z-4| = 1$

Here, the two poles lie outside the circle C.

$\rightarrow \int\limits_{c} \dfrac{dz}{z^3(z+4)} = 0$

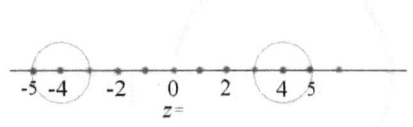

(iii) $|z-4| = 1$

Example 30

Find the value of the integral

$$\int_C \frac{dz}{z^2 e^z}$$

Where the circle C has origin at $z = 0$, with radius $= 1$.

Solution

Here, we a 2nd order pole at $z = 0$

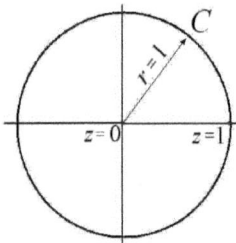

$$\frac{1}{z^2 e^z} = \frac{1}{z^2} e^z$$

$$= \frac{1}{z^2} \left(1 - z + \frac{z^2}{2!} - \frac{z^3}{3!} \right)$$

$$= \frac{1}{z^2} - \frac{1}{z} + \Phi(z)$$

We have stopped at $\frac{1}{z}$ since there no poles beyond that term, which means that $\Phi(z)$ is regular at $z = 0$.

Hence, the Residue $= -1$, the numerator of the term $\frac{1}{z}$.

Therefore,

$$\int_C \frac{dz}{z^2 e^z} = 2\pi i (-1)$$

110

$$= -2\pi i$$

Check using the Cauchy's Theorem of residues, we get:

$$\text{Residue} = \frac{1}{(2-1)!}\left[\frac{d}{dz}\left(\frac{1}{z^2}\right)\right]_{z=0} = -1$$

Example 31

Find the value of the integral

$$\int_0^\infty \frac{dx}{(x^2+1)^2}$$

Solution

$$\int_0^\infty \frac{dx}{(x^2+1)^2} = \frac{1}{2}\int_{-\infty}^\infty \frac{dx}{(x^2+1)^2}$$

This follows from the fact that the integral in z-plane is an even function of x.

Now, consider

$$\int_C \frac{dz}{(z^2+1)^2}$$

Where, C is the contour consisting of the real part of the x-axis, extending from –R to +R and semicircle C, in the upper half of the z-plane, centered at origin with radius R.

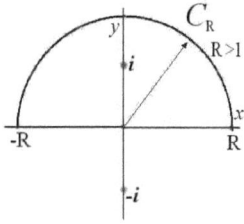

When R > 1, hence

$$\int_C \frac{dz}{(z^2+1)^2} = \int_{-R}^R \frac{dx}{(x^2+1)^2} + \int_{C_R} \frac{dz}{(z^2+1)^2}$$
$$= 2\pi i\,(\text{Residue at } z = i)$$

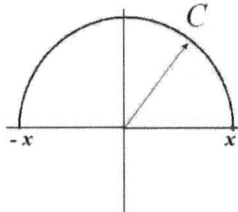

We now calculate the residue at $z = R$.

$$\frac{1}{(z^2+1)^2} = \frac{1}{(z-i)^2(z+i)^2}$$

Residue at $z = i$:

$$= \frac{1}{1!}\left[\frac{d}{dz}\left(\frac{1}{(z+i)^2}\right)\right]_{z=i}$$

$$= \frac{-2}{(i+i)^3}$$

$$= \frac{-2}{8i^3} = \frac{1}{4i}$$

Residue at $z = -i$: Here, the pole lies out the contour, therefore, residue vanishes.

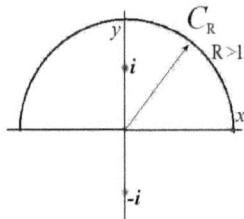

Therefore, the value of the integral is

$$\int_C \frac{dz}{(z^2+1)^2} = 2\pi i \frac{1}{4i} = \frac{\pi}{2}$$

Now, consider the integral

$$\int_{C_R} \frac{dz}{(z^2+1)^2}$$

We will the already proven property of complex variables that the modulus of summation is equal or less than the summation of the moduli, such that

$$|z_1 + z_2| \leq |z_1| + |z_2|$$
And
$$|z_1 + z_2| \geq |z_1| - |z_2|$$

Therefore,

$$\left| \int_{C_R} \frac{dz}{(z^2+1)^2} \right| \leq \int_{C_R} \left| \frac{dz}{(z^2+1)^2} \right|$$

$$< \int_{C_R} \frac{|dz|}{(R^2-1)^2} = \frac{\pi R}{(R^2-1)^2}$$

Therefore,

$$\int_{-\infty}^{\infty} \frac{dx}{(x^2+1)^2} = \frac{\pi}{2}$$

And

$$\int_{0}^{\infty} \frac{dx}{(x^2+1)^2} = \frac{\pi}{4}$$

Another method of solution

We use the polar coordinates to evaluate the integral as follows.

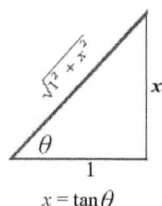

$x = \tan \theta$

Let $x = \tan \theta$

$dx = \sec^2 \theta \, d\theta$

$$\int_{0}^{\infty} \frac{dx}{(x^2+1)^2} = \int \frac{\sec^2\theta \; d\theta}{\sec^4\theta}$$

$$= \int_{\theta=0}^{\theta=\pi/2} \cos^2\theta \; d\theta$$

Decompose the squared cosine using trigonometric rules, we get:

$$= \int_{\theta=0}^{\theta=\pi/2} \frac{1}{2}(1+\cos 2\theta) \; d\theta$$

$$= \frac{1}{2}[\theta]_0^{\pi/2} + \int_{\theta=0}^{\theta=\pi/2} \frac{1}{4}\cos 2\theta \; d(2\theta)$$

$$= \left[\frac{1}{2}\theta + \frac{1}{4}\sin 2\theta\right]_{\theta=0}^{\theta=\pi/2}$$

$$= \frac{\pi}{4} + 0$$

Therefore, using the symmetry of the integral around the x-axis, we get

$$\int_{-\infty}^{\infty} \frac{dx}{(x^2+1)^2} = 2\int_{0}^{\infty} \frac{dx}{(x^2+1)^2}$$

$$= 2\frac{\pi}{4}$$

$$= \frac{\pi}{2}$$

$$\text{Residue} = \frac{1}{(2-1)!}\left[\frac{d}{dz}\left(\frac{1}{z^2}\right)\right]_{z=0} = -1$$

Example 32

Find the value of the integral

$$\int_{0}^{\infty} \frac{\cos x \; dx}{x^2+1}$$

Solution

$$\int_0^\infty \frac{\cos x \ dx}{x^2+1} = \frac{1}{2}\int_{-\infty}^\infty \frac{\cos x \ dx}{x^2+1}$$

Consider the integral

$$\int_C \frac{e^{iz} \ dz}{z^2+1} = \int_{-R}^R \frac{e^{ix} \ dx}{x^2+1} + \int_{C_R} \frac{e^{iz} \ dz}{z^2+1}$$
$$= 2\pi i \ (\text{Residue a } z=i\ ) \ ....(7.15)$$

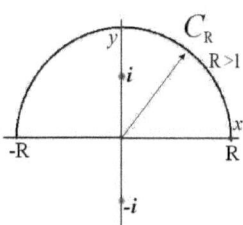

We could write the integrand $\dfrac{e^{iz}}{z^2+1}$ as follows

$$\frac{e^{iz}}{z^2+1} = \frac{e^{iz}}{(z+i)(z-i)}$$

Therefore, we have residue at $z=i$, such that

$$\lim_{z\to i}(z-i)\frac{e^{iz}}{(z+i)(z-i)} = \frac{e^{i^2}}{2i}$$
$$= \frac{e^{-1}}{2i}$$
$$= \frac{1}{2ie} \ ....(7.16)$$

Therefore, equation (1) gives

$$\int_C \frac{e^{iz} \ dz}{z^2+1} = 2\pi i \cdot \frac{1}{2ie}$$
$$= \frac{\pi}{e} \ ....(7.17)$$

We will use the rule of moduli of complex variables as follows

$$\left| \int_C \frac{e^{iz} \ dz}{z^2+1} \right| \le \int_C \frac{|e^{iz}|}{|z^2+1|} |dz|$$

115

And

$$\left|e^{iz}\right| = \left|e^{i(x+iy)}\right|$$
$$= \left|e^{ix}\right| \ e^{-y}$$

$$= \left|\cos x + i\sin x\right| \ e^{-y}$$

Therefore,

$$\left|e^{iz}\right| = \left|\sqrt{\cos^2 x + \sin^2 x}\right| \ e^{-y}$$

$$= 1. \ e^{-y} < 1, \ \text{since } y > 0$$

Therefore,

$$\left|\int_{C_R} \frac{e^{iz} \ dz}{z^2 + 1}\right| < \int_R \frac{e^{-y}}{R^2 + 1} \ |dz| \rightarrow \text{as } R \rightarrow \infty$$

And

$$\int_{-R}^{R} \frac{e^{ix} \ dx}{x^2 + 1} = \frac{\pi}{e}$$

$$= \int_{-\infty}^{\infty} \frac{\cos x + i\sin x}{x^2 + 1} dx$$

Equating the real parts, we get

$$\int_{-\infty}^{\infty} \frac{\cos x}{x^2 + 1} dx = \frac{\pi}{e}$$

And

$$\int_{0}^{\infty} \frac{\cos x \ dx}{x^2 + 1} = \frac{\pi}{2 \ e}$$

7.4. Cauchy's Integral formula

Let $f(z)$ be a function regular in a domain D.
Let (a) be any point of the domain.

Then,

$$f(a) = \frac{1}{2\pi \ i} \int_C \frac{f(z)}{z-a} dz$$

where, C is any closed domain surrounding a and lying in D.
Draw a circle γ, centered at the vector a, with radius ρ lying inside C.

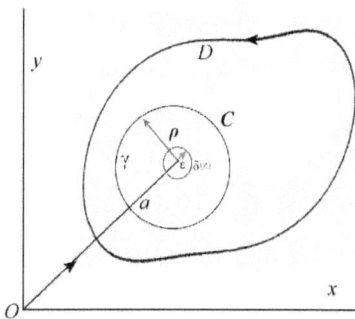

Hence, according to Cauchy's Theorem, we get

$$\int_C \frac{f(z)}{z-a} dz = \int_\gamma \frac{f(z)}{z-a} dz$$

$$= \int_\gamma \frac{f(z)-f(a)}{z-a} dz - \int_\gamma \frac{f(a)}{z-a} dz \dots\dots\dots\dots\dots\dots(7.18)$$

Now, the last integral has the value of

$$\int_\gamma \frac{f(a)}{z-a} dz = 2\pi i \cdot f(a)$$

Where, by hypothesis, $f(a)$ is regular at $z = a$.

Hence, $f(a)$ has derivative at $z = a$ and is, therefore, continuous at $z = a$. The continuity property implies $f(z) \rightarrow f(a)$ as $z = a$.

Hence, given the small distance; ε, around $z = a$, there exists $\delta = \delta(\varepsilon)$, i.e., dependent on ε, such that,

$|f(z) - f(a)| < \varepsilon$

For all z satisfying

$|z - a| < \delta$.

$$\left| \int_\gamma \frac{f(z) - f(a)}{z - a} dz \right| \leq \left| \frac{f(z) - f(a)}{z - a} \right| \left| dz \right|$$

$$< \frac{\varepsilon}{|\rho|} \left| dz \right|$$

$$= \frac{\varepsilon}{|\rho|} \ 2\pi\rho$$

$$= 2\pi \ \varepsilon$$

Hence, the integration the integration on the left of equation (7.17) is zero

$$\int_C \frac{f(z)}{z - a} dz = \int_\gamma \frac{f(z) - f(a)}{z - a} dz - \int_\gamma \frac{f(a)}{z - a} dz$$

Therefore,

$$f(a) = \frac{1}{2\pi \ i} \int_\gamma \frac{f(z)}{z - a} dz$$

CHAPTER 8

TAYLOR'S AND LAURENT'S EXPANSIONS

Let $f(z)$ be a regular function in a circular domain, center around a, radius R. $f(z)$ can be expanded in the neighborhood of a as follows

$$f(z) = A_0 + A_1(z-a) + A_2(z-a)^2 + ...$$

i.e.,

$$f(z) = \sum_{n=0}^{\infty} A_n(z-a)^n$$

The expansion $f(z)$ in powers of $(z-a)$ implies that

$$A_n = \frac{1}{2\pi i} \int_C \frac{f(z)}{(z-a)^{n+1}} dz$$

Or,

$$A_n = \frac{f^n(a)}{n!}$$

And hence,

$$f(z) = f(a + \overline{z-a})$$

$$= f(a) + (z\text{-}a)f'(a) + \frac{(z-a)^2}{2!} f''(a) + .. + \frac{(z-a)^n}{n!} f^n(a)$$

Next, consider the case of an annulus bounded by two circles of radii R_1 and R_2.

Let $f(z)$ be regular function in this annulus. According to Laurent's expansion, $f(z)$ can be expanded in the form:

$$f(z) = \sum_{n=-\infty}^{\infty} A_n(z-a)^n$$

where,

$$A_n = \frac{1}{2\pi i} \int_C \frac{f(z)}{(z-a)^{n+1}} dz$$

And C is any circle of center a and radius lies in between R_1 and R_2.